Thinking Like an Island

Thinking Like an Island

Navigating a Sustainable Future in Hawai'i

JENNIFER CHIRICO AND GREGORY S. FARLEY

Aloha Shan!
Thank you for all
of your contributions
to this field that
positively benefit the
world. Mahalo!

University of Hawai'i Press   *Honolulu*

Printed in the United States of America

23 22 21 20 19 18   6 5 4 3 2 1

Library of Congress Cataloging-in-Publication Data
    Thinking like an island : navigating a sustainable future in Hawai'i / [edited by] Jennifer Chirico and Gregory S. Farley.
        pages cm
    Includes bibliographical references and index.
    ISBN 978-0-8248-4761-6 (hardcover : alk. paper)
    1. Sustainable development—Hawaii.   2. Sustainability—Hawaii.   I. Chirico, Jennifer, editor.
II. Farley, Gregory S., editor.
    HC107.H33E585 2015
    338.9969'07—dc23

                                                                                    2014038341

ISBN 978-0-8248-7666-1 (pbk.)

University of Hawai'i Press books are printed on acid-free
paper and meet the guidelines for permanence and
durability of the Council on Library Resources.

Designed by Milenda Lee

Contents

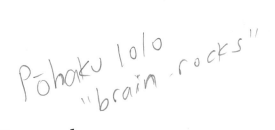

Pōhaku lolo
"brain rocks"

Foreword

RAMSAY REMIGIUS MAHEALANI TAUM

I ka nānā no a ʻike
Learn by observing

Na pōhaku
"the stones"
ʻike -knowledge

As I read through the chapters within, I was transported back to one of the many Saturday morning "talk story" sessions that I was privileged to have with Aunty Pilahi Paki, a well-known teacher and advocate of Hawaiian language and a "keeper of Hawaiʻi's secrets." As I read about the advances in renewable energy technologies and accessing wave energy as a preferred source of energy in the future, I recalled a particular day when I asked her about *na pōhaku* (the stones) that were located around her *hale* (home). "These are my *pōhaku lolo* . . . my brain rocks," she said. "They are like my encyclopedia. Each stone contains the *ʻike* (knowledge) of our *kupuna* (elders). Whenever I need information regardless of whether it is about agriculture, aquaculture, or astronomy, all I have to do is pick up and hold one of these *pōhaku* in my lap."

She went on to explain that "our ancestors had many schools of knowledge and were aware of many sciences that modern scientists are only now becoming aware of, or are slowly beginning to acknowledge and accept." She looked ahead, saying, "There will be a time when you will see your *kupuna*'s knowledge being applied in technologies that will change the way we live in the world." More important, she also predicted that "there will be a time when the world will be in turmoil and Hawaiʻi will play an important role" in helping to bring about healing to an ailing planet.

Reflecting on these lessons and predictions, I smile as I touch the black stones sitting upon the desk in front of me: my iPhone and iPad. Like Aunty Pilahi before me, I can access information from all over the world by holding either of them in my lap. Unlike her, however, I require a monitor or screen to access the information, and my devices need to be recharged while her stones did not. So close and yet so far!

I believe *Thinking Like an Island* qualifies as one of those gifts that Aunty Pilahi was referring to as it offers thoughts and perspectives on sustainable practices and approaches that are appropriately place based and culturally based. What is particularly refreshing about what the authors offer is the acknowledgment that there may be benefits to be gained by understanding the differences between an island worldview and a continental one.

Polynesian seafarers, for instance, settled thousands of islands in the Pacific centuries before their European counterparts had ventured across the Atlantic. Despite the fact that they had successfully traversed 800,000 square miles of ocean, without the aid of modern navigational tools, and established sustainable societies as far north as Hawai'i, the most remote inhabited place on the planet, the latter "discoveries" of European adventurers and explorers would serve as the cornerstone upon which the future understanding and knowledge about the Pacific, Polynesia, and Hawai'i would be built. Rather than build upon the knowledge of what was known, development of the Pacific as we know it today was largely shaped by foreign interests and foreign ideas.

This same Eurocentric, continental worldview would be the source of inspiration and guidance for the development of everything in modern Hawai'i as well. Systems of governance, urban planning, economic development, education, agriculture, and resource management were all influenced by continental thinking rather than island thinking. Island communities and ecosystems were expected to adopt as well as adapt to continental practices and policies, whether they were relevant or appropriate or not.

Thinking Like an Island effectively questions conventional wisdom and approaches to sustainable living by exploring and revealing the benefits of acknowledging local, place-based practices that are as relevant today as they were for ancient Hawaiians. The authors also effectively discuss the tensions inherent in developing, executing, and managing systems to feed, house, and educate a growing population confined to a finite space in the middle of the sea, while maintaining a healthy ecosystem and respecting the values and spirit that are uniquely Hawai'i.

There are lessons to be learned and shared from our experience living on islands and from *Thinking Like an Island*. Given that Earth is more like an island than it is a continent, who better to learn from than those who have successfully lived on islands for thousands of years? *Thinking Like an Island* offers lessons that might serve our island Earth community as much as it will our community in Hawaiʻi.

Ka wa ma mua, ka wa ma hope—the future is in the past

INTRODUCTION

JENNIFER CHIRICO AND GREGORY S. FARLEY

Mālama pono, mālama 'āina, mālama 'ohana
Be a steward of righteousness, care for the land,   .
    take care of your family and community

Sebastian Junger's 1997 best seller *The Perfect Storm* describes the events that led to the loss of the Massachusetts fishing vessel *Andrea Gail:* the economics of offshore fishing, the risks inherent in the enterprise, and the traditional New England culture that continues to encourage, and even romanticize, working at sea. Half a globe away, the same themes are at work in a different way: in Hawai'i, a "perfect storm" of economic need, resource opportunity, and a resurgent culture has led Hawaiians to forge a new, more sustainable future for the world's most geographically isolated landmass. Beset by the nation's highest energy prices, a food-supply system dependent upon long-distance transportation, and the long-term ecological effects of colonial resource exploitation, Hawai'i could remain utterly dependent upon the US mainland and other parts of the world for its every need—but it would pay an ever-increasing cost for that dependence. Instead, buoyed by opportunity, desire, practicality, and the remembrance of native Hawaiian culture and practices, the fiftieth state is becoming a sustainability showcase: a model for sustainable living that is also applicable to isolated communities worldwide. As the authors in this volume illustrate, it is possible to grow food, generate energy, build homes, and educate the population while preserving and drawing strength from the Hawaiian Islands' unique native culture.

Culture emerged as a strong theme throughout this book. What began as a collection of case studies in energy, building science, food production, and resource management began to echo with a steady drumbeat: many authors drew Hawaiian culture into their case studies. It became impossible to overstate the importance of culture as Hawai'i addresses its critical island sustainability issues.

Hawaiians are fortunate in this regard—they both remember their culture in a very immediate way and are experiencing an ongoing renaissance in native Hawaiian culture that began in the 1970s and 1980s. One of the last places to experience the influence of colonialism, Hawai'i was home to a robust and unique culture that tied people to place, and place to people. That experience, coupled with the exquisite Hawaiian memory for genealogies and events, preserved in *mele,* or chants, means that Hawaiians have a strong remembrance of their precolonial past. Historical Hawaiian culture was well fitted to the islands it depended upon, and its memory is reshaping agricultural, cultural, and other practices to reestablish a connection to the land. It is not a small surprise that Olomana's popular 1976 song "Ku'u Home O Kahalu'u," an anthem of the Hawaiian renaissance, sings of a memory of place—and personal involvement with it.

We regard sustainability as a fundamental change, a paradigm shift, in the way that humans view the world. For the last several hundred years, industrial nations have focused primarily on economic growth: How much profit was made? How many items were produced? How many items were consumed? The end result—a high gross domestic product—represented success and was considered the ultimate benefit to society. This concept, however, failed to include assessments of natural resource depletion, social and financial inequality, or cultural loss. Instead, the economic model of the nineteenth and twentieth centuries left a legacy of despoiled environments and displaced communities. This model has made clear that a short-term focus on profits cannot be adequately sustained unless its fundamental assumptions are reexamined.

Sustainability requires us to view the world and all of its processes from a systems perspective: everything is interconnected, and everything is interdependent. Long-term economic health depends upon the health of ecosystems, which in turn supports the health, the productivity, and the existence of future generations. More and more companies are investing in natural resource management and placing a stronger focus on social responsibility. They are finding that these investments foster greater economic prosperity, while also protecting our environment, respecting native cultures, and providing a more equitable foundation so that we may meet our current needs while also protecting the earth for future generations.

Aldo Leopold's 1949 essay "Thinking Like a Mountain," presented in *A Sand County Almanac*, provides an excellent example of the importance of working toward sustainability using a systems perspective. Leopold eloquently describes his experiences working to increase deer population by hunting wolves. It was commonly believed that if all of the wolves were killed, there would be more deer for hunting. In Leopold's essay, he writes about his experience killing a wolf. While looking into the wolf's eyes as it was dying, he came to the realization that fewer wolves did not necessarily lead to a "hunter's paradise." Instead, when the wolves were killed, the deer proliferated. Even though there were more deer for hunting, the greater number of hungry deer ate all of the vegetation, leaving the mountain a wasteland. The message in Leopold's essay is on the importance of valuing the entire system in nature, rather than just the parts. His essay shows how multiscalar relationships are significantly affected by human decisions, which is an important component of strong sustainability. "Thinking like a mountain is thinking about human values as time-sensitive and as produced by specific processes and dynamics that unfold on identifiable scales" (Norton 2005, 230). Islands, like mountains, "think" on many different scales and time, and it is essential for humans to recognize these spatial and temporal aspects of ecosystems.

This perspective is paramount to small, remote islands. Today, most islands depend upon other faraway places for food and energy. Many island economies depend on tourism as their primary economic means, yet the environmental footprint of tourists can be much larger than those of average residents. The Hawaiian Islands offer an excellent example of the issues that remote islands face. Hawai'i is over 90 percent dependent on fossil fuels from other countries for its energy needs (Hawai'i Clean Energy Initiative 2014). Over 85 percent of the food in Hawai'i is imported from other locations (Hawai'i Department of Agriculture 2008). Everything that enters the islands must be disposed of in some way, and Hawai'i's landfills are at capacity with limited land for expansion. Fresh water in Hawai'i is also limited, and much of it is diverted to large agricultural companies. "Water wars" in Hawai'i bring focus to the question of access to resources: Does fresh water belong to Hawaiians who historically used the water for growing their staple food, taro,[1] or to large-scale agribusiness? Agribusiness interests claim the water they capture feeds crops that provide the foundation for the local economy, while many Native Hawaiians claim that the water from their lands should be diverted back to its original system to revive the dying taro fields.

Historical Hawaiian culture provides a rich alternative model to irrigation-dependent large-scale agriculture, a model that, at its peak, fostered a sustainable population of approximately one million Hawaiians (Dye 1994). Prior to European contact, Hawaiian geography was parceled into *ahupua'a,* which were wedge-shaped land boundaries that extended from mountaintop to sea and included various forms of terrain. Within the *ahupua'a* system, every Hawaiian community had freshwater streams coming from the mountains, community centers in the middle for farming, and oceanside areas for fishing. Most of a community's material needs were met within a single *ahupua'a.* Houses were made of the land, from banana stalks and lava rocks set together with pebbles or sand. Agriculture existed wherever there was topsoil. Banana, coconut, and breadfruit groves provided housing, household items, and food. Hawaiians lived in harmony with the land; they utilized a cyclical process that was aligned with nature: everything was biodegradable, and everything was returned to the land to be used as food for the next cycle (Kane 1997). Maintaining the integrity of, and respect for, Hawaiian culture is a key component of working toward greater sustainability in Hawai'i. Many traditional Hawaiian resource-use practices can help model modern solutions to resource management issues.

This book provides examples of how many of these issues are being addressed in Hawai'i, how successful implementation of sustainable strategies has been achieved, and how cultural sustainability contributes to ecological and economic sustainability. All of the chapters were written by authors who permanently reside and work in Hawai'i. We purposefully chose authors in Hawai'i who were working close to the issues on the ground so that we could deliver an authentic and real account of the sustainability challenges, illustrate how those challenges were overcome, and provide recommendations on strategies for other islands and isolated communities worldwide.

The first section of the book explicitly examines the role and centrality of Hawaiian culture in the success of sustainable strategies. The first chapter describes the integrated nature of the Hawaiian social and agricultural system and how it was displaced by "western" economic models after European contact. Chapter 2 illuminates the challenges of food security in Hawai'i. Chapters 3 and 4 discuss the history of Native Hawaiian sustainable agriculture and how it can be revived to facilitate a movement towards greater food security.

Chapters 5 and 6 draw upon authors' experiences with conservation and the protection of fresh water, a vital issue in modern Hawai'i. These chapters rely more upon technology than culture and demonstrate how natural water systems can provide long-term sustainable approaches for water and wastewater management (chapter 5) and a synthesis of policy and technology in water reclamation and recycling (chapter 6).

Chapters 7 and 8 discuss renewable energy and energy efficiency approaches that have been, or are planned to be, implemented in Hawai'i. Chapter 7 examines the potential for wave energy generation as a renewable energy in fossil-fuel-dependent Hawai'i.[2] Chapter 8 examines green building approaches for implementing more sustainable construction techniques and returns, in a way, to the theme of native Hawaiian culture, as modern building techniques begin to resemble the ways in which precolonial Hawaiians built structures.

The final three chapters assess green tourism (chapter 9), which is a growing and central component of the island economies; a successful sustainability initiative in higher education (chapter 10); and a school garden movement for children (chapter 11). Taken together, these three chapters stress the importance of involving all participants, including consumers and students, in a culture of sustainability.

True change will be realized, and the new, emerging paradigm fully actualized, only when society as a whole shifts its view of the natural world. Islands and isolated communities around the world are facing circumstances that could lead to the "perfect storm." Sea level rise already threatens many island nations and coastal regions; warming climates and acidified oceans threaten the fragile resources upon which island economies depend. Meanwhile, global population continues to grow at an accelerating rate, leaving the earth's finite resources more and more stressed. If the resource exploitation models of the twentieth century do not change, natural resources will inevitably become scarcer. Communities will be faced, at many levels, with a singular choice: adopt more sustainable practices and survive, or continue with "business as usual" and fail.

Fortunately, successful new and old strategies are already available. Examples from Hawai'i show how islands and other isolated communities around the world can forge forward toward a more sustainable future. This book's collection of success stories provides a road map for communities traveling along the path to sustainability. Every community is different, and every challenge requires its own solution; the reader will find no "one-size-fits-all" solution.

However, the authors in this book show that a combination of new technology, rediscovered practices, personal conviction, and courage can both draw from, and help rebuild, strong, vital, sustainable communities.

NOTES

1. Taro, or *kalo,* is the most common Hawaiian staple food (see chapters 3 and 4).

2. We have not attempted a comprehensive review of renewable energy generation in Hawai'i, which has among the highest per capita solar- and wind-generated energy capacities in the world. These are mature technologies; we decided to focus instead on a newer, lesser known technology—wave energy—to explore the possibilities for Hawai'i and elsewhere. We also considered geothermal electricity generation, but since geothermal is being utilized only for Hawai'i Island, and is not a viable option for the other islands, it was withdrawn from the book.

REFERENCES

Dye, T. 1994. "Population Trends in Hawaii before 1778." *Hawaiian Journal of History* 28: 1–20.

Hawai'i Clean Energy Initiative. 2014. Home page. http://www.hawaiicleanenergyinitiative.org/.

Hawai'i Department of Agriculture. 2008. *Food Self-Sufficiency in Hawai'i.* http://hawaii.gov/hdoa/add/White%20Paper%20D14.pdf.

Junger, S. 1997. *The Perfect Storm: A True Story of Men against the Sea.* New York: Norton.

Kane, H. K. 1997. *Ancient Hawaii.* Captain Cook, Hawai'i: Kawainui Press.

Leopold, A. 1949. "Thinking Like a Mountain." In *A Sand County Almanac, and Sketches Here and There,* 129–132. New York: Oxford University Press.

Norton, B. G. 2005. *Sustainability: A Philosophy of Adaptive Ecosystem Management.* Chicago: University of Chicago Press.

ONE

Hawaiian Culture and Its Foundation in Sustainability

SCOTT FISHER

Ua lehulehu a manomano ka 'ikena a ka Hawai'i
Great and numerous is the knowledge of the Hawaiians
MARY KAWENA PUKUI, *'ŌLELO NO'EAU*

As the ancestors of the Hawaiian people moved across western Oceania, they consistently met the challenge of living sustainably on geographically limited islands by forming strong and enduring familial relationships with the lands and ocean that supported them. Over the millennia, this developed into a worldview that required the careful protection of natural resources to perpetuate the life of the community. Hawaiian sustainability was founded upon this island worldview, which places an intrinsic value on the interdependence of all life, with a sense of the sacred permeating the entire natural world. The Hawaiian connection to the land and ocean and their understanding of the sacred nature of this relationship fostered a sustainable culture in Hawai'i for generations.

This chapter describes the foundation of Hawaiian sustainability, suggesting that it primarily consists of three intimately related pillars. It then identifies some of the underlying cultural sources that perpetuated Hawaiian sustainability prior to European contact in 1778, followed by an analysis of the historical context in Hawai'i in the eighteenth and nineteenth centuries, which eroded this foundational worldview to the point that it survived only in remnants. It explores the reclamation of certain aspects of Hawaiian culture during the cultural renaissance in the latter half of the twentieth century. I argue that the reassertion of these Hawaiian values can be seen in the context of retrieval, reevaluation, and transformation. Specifically, the efforts to retrieve the island paradigm must be understood in light of contemporary problems and issues in order to transform them in a way that makes them relevant to our modern world. A case study is provided from the Waiheʻe Coastal Dunes and Wetlands Refuge on Maui's North Shore, where I have worked as the manager for the past eight years, to illustrate how these elements of sustainability might find their modern expression. This work is fundamentally an overview of sustainability in Hawaiʻi with reference to the cultural underpinnings of sustainability, its loss that brings us to the unsustainable present, and the attempts to reclaim Hawaiian history and traditions and the relation to the land that has sustained Hawaiian culture for millennia. Implementing these ideas has taken on a sense of priority and urgency in recent decades. As Chun (2009, vii) states, "Cultural revival and identification have gone beyond academic and intellectual arguments to a reality in communities and families and are now part of the political landscape of the islands."

THE FOUNDATION OF HAWAIIAN SUSTAINABILITY

Sometime between 300 and 600 AD, the first humans arrived on Hawaiian shores from southern Polynesia, probably from either Tahiti or the Marquesas Islands. The ancestors of this early Hawaiian community migrated eastward across much of southern Oceania over the previous millennia, populating both high islands and atolls during their voyages. This pattern of movement across virtually the entire Pacific Ocean over the course of thousands of years helped to cultivate in the Polynesian people a cultural worldview that placed islands, and the oceans that connected them, at the center of their cultural paradigm.

Like their Polynesian cousins, the Hawaiian people developed their own cultural norms and values but share an awareness of the island worldview with their relatives spread out across Melanesia, Micronesia, and Polynesia. This worldview contains a number of related features and is moderated by both individual cultural expressions and the unique location and setting of their island home. In general, the island worldview stands in contrast to the western worldview in its understanding of the inherent limitations of island ecosystems. This awareness of the bounded space of the island environment frequently expressed itself in the establishment of well-defined laws that protected natural resources, particularly those necessary for human welfare, from overexploitation. Perhaps because of its isolation, the early Hawaiian community adopted a number of cultural attitudes and behaviors that strengthened relations laterally within and among community members while also expressing a sacred relationship with the natural world.

Anthropologists have identified several distinct phases of Hawaiian cultural development, each demonstrating an evolving relationship both with the natural world and within and between familial relations. These phases coincided with an increasing awareness of the extent and limitations of the endemic resources available to the Hawaiian population and the intensification of agrarian practices that fostered settlement in environmentally marginal areas (particularly areas with low annual rainfalls). While the Hawaiian community initiated significant changes to the native ecosystems, it also developed a society that met the needs of the local population while also expressing a high degree of ecological sensitivity.

In precontact Hawai'i, the community sustained itself through the establishment of reciprocal and familial relationships between humans and the natural world, as well as among the various human communities. In this regard, Hawaiians asserted their right to appropriately harvest those items critical to human welfare while also acknowledging an obligation to ensure the long-term health of the ecosystems upon which all life depended. The three pillars of sustainability in Hawai'i prior to the arrival of Europeans involved three closely related concepts. The first of these derived from their cultural evolution on island ecosystems, which are described here as an "island worldview." The second pillar derived from the Hawaiian religious connection to the natural world and manifested as an awareness of the community's familial relationship with natural phenomena. The third pillar of sustainability emerged out of the sense of interdependence, particularly between the upper or *ali'i* class and the commoners (*maka'āinana*), as well the between the people and land that supported them.

Each of these three elements contributed to making precontact Hawai'i a sustainable and thriving community, capable of feeding its relatively large population without relying on external imports. While Hawai'i once demonstrated its sustainability based on an expression of communal interdependence, the Hawaiian Islands today have largely devolved into a society dependent on the importation of the majority of its food and other vital resources (see, e.g., chapter 2 of this volume). The following section presents the cultural underpinnings that promoted sustainability, followed by an overview of Hawaiian history, which brought us to this condition.

Mana — expression of nonordinary power, capacity, or capability.

KE KUMU: THE SOURCE OF HAWAIIAN SUSTAINABILITY

The Hawaiian concept of *mana* stands out as one of the most fundamental concepts in the Hawaiian worldview and serves as a gateway to understanding numerous dimensions of the Hawaiian culture. *Mana* is most often translated as an expression of supernatural power, or an expression of nonordinary power, capacity, or capability. Humans, in particular, primarily derived their *mana* through descent from the gods or through transmission by the community at large. However, both living and nonliving objects contained a measure of *mana*, derived primarily from their relative position in the cosmogenic lineage. As Valeri (1985, 99) defines it, "*Mana* is the efficacy of a working 'fellowship,' it is substantialized, probably because it is viewed as a quality that circulates, that can be moved from one term of the fellowship to another." While *mana* may be defined as supernatural, it is primarily expressed through the familial relations between and among one's 'ohana, or extended family, and the 'āina, or land, that sustained them.

Kapu, or laws (literally, "sacred"), were established in order to preserve or protect the community's *mana*, and primarily ensured that it was not stolen or otherwise misappropriated. The *kapu* are specifically designed to ensure the purity of the *mana*, which adheres in an object. As Valeri (1985, 92) points out, "In Hawai'i, purity is first and foremost a positive quality: it is the mark of a close instantiation of the divine . . . different degrees of purity, of *mana*, must be associated in a way that preserves their hierarchical distinction." In this regard, the series of prohibitions that can collectively be called *kapu* ultimately served as a barrier against the defilement (*haumia*) of *mana*. The essence of the island worldview as it relates to *mana* and *kapu* derives from the essential role the community played in protecting the *mana* of the land through the establishment of *kapu*. These *kapu* ensured that items that sustained the community were harvested at the proper times and in appropriate quantities. Through this reliance on proscriptions and prohibitions, which came about through careful observations and a deep sense of communal interdependence, the precontact Hawaiian community achieved one of the most fundamental aspects of sustainability.

However, this sense of interdependence extended beyond the human realm into the divine realm, the ultimate repository and source of *mana*. This is most clearly seen in one's ancestral spirits, or 'aumākua, which were frequently empowered or established as familial deities in a ritual known as *kaku'ai*. This ritual involved the communal imparting of *mana* on a deceased relative in order to transfigure this individual into a natural or totemic spirit. As Hawaiian scholar Mary Kawena Pukui (1979, 35) explains: "The Hawaiians lived within the close relationships of the ʻohana (family or family clan); the ʻaumākua remained members of the clan . . . . with one's ʻaumākua, a human-to-spirit communication was possible." However, the ʻaumākua of a family were always embodied within a specific animal or natural phenomenon, in such bodily forms as a shark, turtle, owl, or hawk, as well as in volcanoes or lightning.

ʻAumākua had a primary function of ensuring the protection of the family, particularly at death, when an individual would be led to the place where she or he would rejoin their ancestors. ʻAumākua also served a vital role as a link to the natural world. In this regard, the interdependence fostered by the family's shared mana and the kapu they observed was elevated to both the natural and spiritual realms.

On island communities, where resources are intrinsically limited or bounded, practices recognized as fostering "sustainability" are, from a Hawaiian, or island, worldview, upheld by the establishment of kapu, which preserve a sense of the sanctity of the mana found in or around that object. ʻAumākua, in turn, fostered this sense of sustainability by enhancing the community's relationship with one's ancestors and the natural world. As Pukui (1979, 220) points out, "Hawaiians of the past lived in an interdependent society based on a subsistence economy." This interdependency fostered a sense of reliance on one's community, as well as the bounty of the natural environment from which all life ultimately derived.

This sense of community interdependence, particularly through its embodiment in the extended family, or ʻohana, expressed itself in the primary land divisions of Hawaiʻi prior to the arrival of Europeans in 1778. The most fundamental land division, the island or mokupuni, was subdivided into districts, or moku, which radiated downward from the summit of the mountain to include the ocean, and in particular the area where the color transition of the nearshore light blue ocean to the dark purple (lipo) of the open ocean.[1] Each moku was then further subdivided into ahupuaʻa of varying sizes, largely dependent on the resources available within that ahupuaʻa (Cordy 2000).

The ahupuaʻa comprised the most basic unit of land within which an individual had the right to access resources. While the most common sites of resource utilization occurred within the wao kanaka, or the human realm, an individual might occasionally visit the wao akua, or the divine realm in the upper forests, in order to harvest less frequently utilized material, such as logs for home building or material for making the cordage rope used to lash canoes and homes together. Additionally, marine resources were available exclusively to the residents of the *ahupua'a* in the nearshore waters, while pelagic resources, those found in the *lipo*, or dark purple ocean, were accessible to anyone on the island based on their needs, talents, and skills. The chief of an *ahupua'a*, known as the *konohiki*, utilized a number of strategies to ensure the availability and fair distribution of resources. As Cordy (2000, 39) describes, "The lords of the *ahupua'a* could restrict communal access to the *ahupua'a* fisheries by placing specific fish under *kapu* for a period of time." Numerous improvements to the land, in the form of high-yielding garden plots, wet taro patches (*lo'i*), and fishponds, ensured an abundance of food resources on an *ahupua'a*, while the responsibility for ensuring its fair distribution rested with the *konohiki*. Another function of the *konohiki* lay in ensuring that all work resulted in the common good. Handy and Handy (1991, 307) point out that "*laulima* (many hands) was the Hawaiian word applied to cooperative enterprise . . . in the building of taro *lo'i* and irrigation ditches, and of fishponds, it was organized and supervised by *konohiki* carrying out orders of their *ali'i*."[2]

Ahupua'a varied in size according to the resources available and the ease with which those resources could be reasonably distributed among local residents. The vast majority of the population, perhaps up to 95 percent, were *maka'āinana*, or commoners in the precontact period (Cordy 2000, 51). While the *maka'āinana* are occasionally compared with serfs in Western Europe, the reality is far more complex. For example, *maka'āinana* were welcome to leave the land they worked if they felt mistreated by the *ali'i*.[3] In one of the last battles on Maui, the *maka'āinana* of Kula, a *moku* on Maui, led an uprising against an *ali'i* who had consistently maltreated the commoners by stealing their pigs. In what is known as *Ke Kaua o Ka 'Aihue Pua'a o Kūkeawe* (Battle of the Stolen Pigs of Ku Keawe), Hoapili, the chief of Maui at the time, came to the assistance of the *maka'āinana* against the *ali'i*. According to Handy and Handy (1991, 63), the *ali'i nui* (high chief) "was a medium in whom was vested divine power and authority . . . but this investment . . . was instrumental in providing only a channeling of power and authority, not a vested right . . . the instances in which an *ali'i* nui was rejected and even killed because of his abuse of his role are sufficient proof that it was not personal authority but trusteeship that established right." The degree to which the *ali'i* understood and expressed the intimate connection between the people and the land, combined with how effectively she or he fostered and promoted communal interdependence, became the measure of the *ali'i*'s success.

Cordy (2000, 31, 50) notes that, in the early postcontact period, Maui had just over two hundred *ahupua'a* in twelve *moku,* and Hawai'i Island had over six hundred *ahupua'a* in six *moku.* A typical *ahupua'a* would have held between one hundred and two hundred inhabitants, with larger *ahupua'a* comprising between four hundred and five hundred residents. Waipi'o Valley on the island of Hawai'i had as many as 2,600 residents in precontact times, largely due to its abundant taro cultivation. Although Hawaiian demographics are a hotly debated topic, Cordy suggests that Hawai'i Island may have had as many as 106,000 residents. It seems reasonable to hypothesize that each of the other major high islands may have had populations of similar size, with the islands of Kaho'olawe, Ni'ihau, and Lana'i having proportionately fewer residents due to their smaller size and somewhat more limited natural resources. In any given resource district, water represented the primary resource limitation. However, substantial populations inhabited the dry *kona,* or leeward, coasts of Maui, as illustrated by the large number of archaeological features in this area.

While this area is very likely much drier today than it was prior to the introduction of ungulates (in particular, goats, deer, and the Russian boar), even before the advent of these alien species intimate ecological and environmental knowledge was critical to the success of the communities. The Hawaiian proverb *"Hahai no ka ua i ka ululā'au,"* which means that the rain follows the forest, suggests an understanding that preserving and protecting the forests remains a crucial element in sustaining the human population (Pukui 1983, 50; Handy and Handy 1991). In these remote, drier areas, structures appear to have been constructed that functioned to slow down the velocity of the water, acting as a mechanism for trapping nutrient-laden sediments. By regulating the velocity of water flow and capturing these sediments, crops such as *'uala,* or sweet potato, could have been cultivated in an area otherwise unsuitable for Polynesian-style wet taro cultivation. Additionally, where terrestrial springs were unavailable, the community was able to utilize freshwater springs that emerged in the ocean by capturing fresh water in gourds. This intimate knowledge of the land, gained through careful observation, stands as a clear example of the importance of the island worldview in fostering sustainability in an otherwise relatively desolate area. Because islands are limited in geographic scope, knowing how to maximize the potential yield in a resource limited area was vital to the perpetuation of a thriving community.

Water, then, operated as both the means of promoting sustainability and the gateway through which the community flourished. As Handy and Handy

(1991, 64) note, "Fresh water as a life-giver was not to the Hawaiians merely a physical element"; it had a far more intimate connotation. In prayers of thanks and invocations used in offering the fruits of the land, in prayers chanted when planting, and in prayers for rain, the "Water of Life of Kāne [the Hawaiian deity of fresh water] is referred to over and over again." Water was seen as the primary source of wealth (with the word for water, *wai*, doubled to mean wealth, *waiwai*), as well as laws. While the west tends to see laws in terms of the "the law of the land," Hawaiians understood water as the genesis of the legal code. As a result, *kanawai*, literally "belonging to the waters," expresses the Hawaiian understanding of the basis of the legal code.

One of the clearest disparities between the precontact allocation of water and the largely unsustainable modern use of water derives from where the supply of water ultimately ends. The precontact Hawaiian community mastered the art of water utilization and transportation through the construction of *'auwai*, or aqueducts, used to feed their *lo'i*. However, this water would have remained within the confines of the *ahupua'a* simply because access to cool, clean water made the difference between wealth (*waiwai*) and impoverishment (*hune*), and ultimately life and death.

While the ease of available water allowed for the community to flourish, a reliance on a readily abundant quantity of water was not seen as an absolute barrier to human habitation. Rather, through the ingenuity that comes from a culture that evolved for millennia on islands, an intimate knowledge of and familial connection to the environment and the natural world that shaped the land, and an interdependent community, the Hawaiian people flourished as stewards of a sustainable island community.[4]

HISTORICAL OVERVIEW: FROM INTERDEPENDENT SUSTAINABILITY TO DEPENDENCE ON EXTERNAL COMMODITIES

Although demographers disagree on the size of the Hawaiian population on the eve of Captain James Cook's arrival in 1778, the reality of Hawai'i's isolation, coupled with its relatively extensive political and social evolution, suggests a high degree of interdependent sustainability. The arrival of foreigners in Hawai'i in the late eighteenth century inaugurated an era of profound changes, which ultimately culminated in the loss of Hawaiian political and economic independence. In large part, these changes were the result of intensified warfare with the arrival of foreign weapons, particularly muskets, cannons, and gunboats; foreign diseases, which decimated the indigenous population; and a demographic shift from rural areas to newly established population centers. Socially, the arrival of foreigners in general, and congregational missionaries in particular, inaugurated a rapid transition from an island worldview to a European or western worldview. As described below, the retention of this traditional island worldview provides the basis for a renewed expression of Hawaiian culture and the values that promoted sustainability (cf. McGregor 2007).

Because of its central location in the Pacific, Hawai'i became a favored destination for resupplying ships travelling between the west coast of North America and Asia. Within a short time, however, Hawai'i's large supply of endemic sandalwood trees (*Santalum* spp.) were recognized as a trade commodity in the burgeoning ports of Hong Kong and Macao. Schooners soon arrived with supplies from the US west coast, often in the form of muskets and other arms, trading these items for sandalwood logs and selling the wood primarily for use as incense in China and the Far East. The sandalwood trade had an almost immediate effect on both the intensity and lethality of Hawaiian warfare. With the prevalence of an extensive array of foreign weapons, high chiefs on several islands made increasingly bold moves to expand their authority to adjacent islands. For example, the high chief of Maui, Kahekili, invaded and conquered O'ahu once he had greater access to foreign muskets and a gunboat. Likewise, one decisive factor in King Kamehameha's victory over Maui in 1790 came about through his acquisition of arms, particularly muskets and cannons from the converted gunboat *Fair American* (Daws 1968).

Because of the fear generated by the growing importation of western weapons and armament, chiefs made significant demands on the *maka'āinana* to harvest a dwindling supply of sandalwood in increasingly remote areas. As McGregor (2007, 30) points out, "Periodically, the common people suffered famines that gripped the land as the chiefs gave priority to meeting the needs of the fur and sandalwood traders." The labor involved in harvesting and extracting these sandalwood logs, labor that was taken away from crop production, resulted in a significant reduction in the available food supply and episodic famines in a number of areas across the Hawaiian archipelago. *Maka'āinana* were hit particularly hard as a result of both the intensified warfare and the reduction in a steady food supply.[5] Sustainability in precontact Hawai'i was largely predicated on the interdependence between the *ali'i* chiefly class and the *maka'āinana* commoners. The resulting breakdown of interdependence initiated by the sandalwood trade marked a significant turning point from Hawai'i's precontact sustainability.

In the years following European contact, Hawai'i also experienced a significant drop in its population. For example, by the early 1820s, the population had crashed to just 135,000 individuals (Silva 2004, 24) from a precontact population that likely well exceeded half a million individuals (Daws 1968, 168). While famines and warfare undoubtedly took their toll, foreign diseases, particularly sexually transmitted diseases, cholera, and smallpox, devastated Hawaiians across the archipelago. By the 1870s Hawai'i's population had fallen below 60,000 (168). One observer in the middle of the nineteenth century wrote of this population loss, pointing out that in the town of Waimea, "it can scarcely be said that there is any native population at all. The hill sides and banks of watercourses show for miles the ruins of the 'olden time'-stone walls half sunk in the ground, broken down and covered with grass" (Kenway, qtd. in Daws 1968, 168).

While the population contracted because of various epidemics throughout the nineteenth century, a second blow to Hawaiian sustainability came about through a significant drop in long-term agricultural productivity. The once intensively farmed areas across the archipelago were largely abandoned as people moved from rural areas to the seaport towns across Hawai'i. With the demographic shifts of the nineteenth century, the intimacy of interdependence eroded quickly between the commoners and chiefs, resulting in greater dependence on foreign goods, commodities, and perhaps most problematically, advocacy of a worldview that promoted profits over the well-being of the people. The misfortunes of disease and the demographic shifts that came about in the postcontact period led to numerous attempts by Hawaiians to explain the new reality (Kame'eleihiwa 1992, 80–82). As Silva (2004, 24) points out, "Regardless of how they attempted to explain the mass deaths . . . the ali'i and the maka'āinana grieved over the enormous loss of life and worried about the fate of the lāhui [nation]—on a scale we today can barely imagine."

Both the population crash of the nineteenth century and the demographic shifts that accompanied them illustrate significant episodes that reduced Hawai'i's independence and sustainability. However, these historic events must be understood in the context of the significant changes in the Hawaiian cultural paradigm. Among the most profound changes to affect the familial connection of the Hawaiian people to the land came with the abandonment of the traditional social and religious order in 1819 and the subsequent arrival of Congregationalist (Calvinist) missionaries the following year.

As both Christians and Americans, the missionaries brought with them a western worldview that fostered notions of individuality and the limitless availability of resources for human consumption. The imparting of this western worldview left little room for the acceptance or acknowledgment of a paradigm particularly adapted to island ecosystems. Perhaps more important, the familial connection to the land, which had been a mainstay of the Hawaiian religious worldview, was replaced by a Christian understanding of the human relation to the natural world that humans had divinely ordained dominion over the natural world (cf. Genesis 1:26 and 1:28).[6] While Hawaiians understood the natural world to be physical manifestations of divine agency (*mana*), the Congregationalist missionaries derided this as idol worship and animism (Thurston 1934; Ellis 1963).

In addition to these sociopolitical transformations, significant environmental changes took place in the wake of western contact. The arrival of ungulates, particularly cattle and goats, inaugurated profound alterations to Hawai'i's ecosystems. These animals contributed, and in many cases accelerated, the denuding of large areas of once densely forested areas. With the loss of forest cover, a cascading sequence of soil exposure, erosion, and nearshore marine degradation resulted. Each of these impacts contributed to the degradation of Hawai'i's long-term environmental sustainability.

Between the arrival of Captain James Cook in 1778 and the establishment of the Territory of Hawai'i in 1898, Hawai'i underwent profound changes that severely unraveled the threads of Hawaiian sustainability developed through generations of intimate and familial relationships with the land.[7] First, the western worldview fostered the perception of humans' intrinsic right of dominion over the natural world (particularly as it was expressed in the book of Genesis) and had taken over the endemic island paradigm, which understood the limitations of a bounded and isolated environment. Second, the religious worldview of Christianity, itself an expression of a western worldview, had severely eroded this intimate and sacred connection to the land. Third, the demographic shifts that resulted from both epidemics and migrations to harbor towns profoundly altered the sense of communal interdependence that had fostered a sustainable Hawai'i for generations.

Fortunately, much knowledge was retained by Hawaiian cultural historians who were aware of the need to preserve the accrued wisdom of the Hawaiian way of life. Often, this knowledge was preserved in stories and legends passed down by elders, or *kūpuna*, from generation to generation. However, historians such as nineteenth-century authors and historians David Malo, John Papa 'Ī'ī, and Samuel Kamakau, among others, preserved substantial volumes of wisdom.[8] These sources provided an extensive basis for the recovery, rebirth, and reassertion of Hawaiian cultural values during the Hawaiian renaissance of the late twentieth century.

RECLAIMING A SUSTAINABLE HAWAI'I: THE HAWAIIAN RENAISSANCE AND THE PERPETUATION OF THE HAWAIIAN CULTURE

In the years after the illegal overthrow of the Hawaiian monarchy in 1893 through the middle of the twentieth century, many important aspects of Hawaiian culture were maintained by a number of cultural practitioners. In spite of the efforts to squelch Hawaiian cultural practices, replace the Hawaiian, or island, worldview with a western worldview, and integrate Hawaiians into contemporary American culture, many traditional values and practices remained to foster a renaissance that began after World War II. While the Hawaiian renaissance could be seen in the context of the worldwide push for cultural expression and independence from colonialism in the latter half of the twentieth century, it represents a unique expression of traditional Hawaiian culture, values, and norms.

One of the most important facets of this reassertion of what it means to live as a contemporary Hawaiian in contemporary Hawai'i stems from an ability to straddle both western and Polynesian worldviews. This involves a threefold process of retrieving knowledge and insights passed down through generations (particularly from *kūpuna*), including how one might live attuned to one's environment; reevaluating the norms, values, and practices in light of the contemporary situation; and transforming them to make them relevant to today. Ultimately, this process affords Hawaiians the opportunity to restore the three pillars of sustainability proposed here (an island worldview, a sense of connection to the land, and communal interdependence) in a way that is neither obsolete nor anachronistic. Like any renaissance movement, the reassertion of the Hawaiian culture includes many facets and dimensions. Their underlying and unifying aspects stem from the fact that they are a conscious and coherent expression of one's understanding of traditional Hawaiian values and are conveyed through political, artistic, and religious dimensions, among others.

The efforts of Protect Kaho'olawe 'Ohana (PKO) to halt the military bombing of the Hawaiian island of Kaho'olawe in the early 1970s stand as one of the most effective and coherent expressions of Hawaiian values put into political action.[9] The PKO began as an organization intent on direct action designed to disrupt the US Navy's use of Kahoʻolawe as an impact area. The goals of the PKO were threefold:

- Ensure, through *aloha ʻāina* (love of the land), the proper use of Hawaiʻi's natural resources (peoples, lands, waters, and all that comes willingly from the *ʻāina* [land])
- Perpetuate the historical, cultural, spiritual, and social significance of Kahoʻolawe
- Instill a strong sense of pride in *hoʻohawaiʻi* (being and acting Hawaiian) through knowledge and practice (Ritte and Sawyer 1978, 2)

Although its most immediate goal lay in halting the bombing of Kahoʻolawe and preventing the further destruction of a Hawaiian island that personified the body of the deity Kanaloa, it also looked ahead to a time when it could "replenish the *ʻāina*" with plants indigenous to Hawaiʻi and instill "the value of aloha *ʻāina*" (2).

The PKO inaugurated its direct action through a series of occupations on Kahoʻolawe beginning in January of 1976. By surreptitiously landing activists on Kahoʻolawe at night, the PKO succeeded in bringing attention to both the plight of Kahoʻolawe and the Hawaiian people and served as a forum for reasserting the Hawaiian value of the love of the land. As George Helm, a founding member of the PKO, pointed out, as a Hawaiian his responsibility lay in protecting Kahoʻolawe: "I am a Hawaiian, and I've inherited the soul of my *kūpuna*. It is my moral responsibility to attempt an ending to this desecration of our sacred *ʻāina*" (Ritte and Sawyer 1978, 2). Helm goes on to point out that, "the truth is, there is man and there is environment. One does not supersede the other. The breath in man is the breath of *Papa* (the earth). Man is merely the caretaker of the land that maintains his life and nourishes his soul. Therefore *ʻāina* is sacred" (27).

The actions of the PKO should be seen in the context of an attempt to retrieve traditional Hawaiian values in light of the urgency of redressing the desecration of the island of Kahoʻolawe. Putting these values into action required both critical analysis and coherent transformation. Additionally, Helm's understanding of the values handed down by his elders (*kūpuna*) illustrates his attempt to recover the sense of the sacredness of the land. Access to land and the ability to implement traditional values and practices remain two of the most important aspects of restoring Hawaiʻi to its former sustainability. In the years since the cessation of military training exercises on Kahoʻolawe, the reassertion of the values espoused by the PKO has found its expression in numerous other realms, both on Kahoʻolawe and elsewhere. As McGregor (2007, 268) points out, Molokaʻi PKO members "organized the groups who visited [Kahoʻolawe] to work as an 'ohana, or extended family, in practicing the values of aloha kekahi i kekahi, or love and respect for each other, and laulima, or cooperative work. They organized the work projects in fulfillment of aloha ʻāina and mālama ʻāina, or care for the land."

While the political assertion of Hawaiian values became one tangible aspect of the recovery of the Hawaiian culture, the Hawaiian renaissance included numerous other dimensions as well. In particular, Hawaiian cultural practitioners made great strides toward reviving traditional elements, such as language and dance (hula). For example, the development of the Hawaiian immersion schools in existence today allows the Hawaiian language to be the medium of instruction from preschool through college. The recovery of traditional Hawaiian celestial navigation and sailing methods remains one of the most celebrated components of the Hawaiian cultural renaissance. Beginning in the early 1970s, the Polynesian Voyaging Society sought to revive these skills by relying on master navigators from Micronesia to teach this lost art form. The successful crossing of the double-hulled canoe Hōkūleʻa between Hawaiʻi and Tahiti in 1976 demonstrated that, centuries before European contact, voyages between various parts of Oceania were possible, as Hawaiian oral histories had suggested.

Numerous local communities continue to implement culturally relevant practices that foster a growing sense of community interdependence and sustainability. McGregor highlights the work of the Ke Aupuni Lokahi Enterprise Community on Molokaʻi as an example of an organization dedicated to promoting a vision of healthy communities linked to a nurtured and protected land. As she points out, "The sacred and dependent relationship between the land and the people sustained Molokaʻi for a thousand years, and the vision statement affirms aloha ʻāina as the bedrock value upon which Molokaʻi's economic recovery will be founded" (2007, 299).

The assertion of the value of the traditional culture, through such things as dance, language, sailing, and political action, has had a profoundly uplifting effect on the Hawaiian people. Each of these had in common a desire to retrieve wisdom and insights frequently devalued by proponents of the western worldview (or, in the case of celestial navigation, lost altogether). Through this process, the Hawaiian people today are well on the way to reestablishing the three pillars of Hawaiian sustainability identified in this chapter. However, much work remains to be done. While numerous examples exist across Hawai'i that illustrate the movement towards sustainability as traditionally understood, below is one example of the movement on the island of Maui in the community of Waihe'e.

A CASE STUDY FOR SUSTAINABILITY: THE WAIHE'E COASTAL DUNES AND WETLANDS REFUGE

In July of 2004, the Maui Coastal Land Trust (MCLT), which has since merged with three other land trusts to form the Hawaiian Islands Land Trust (HILT), purchased a 277-acre parcel that had been slated for development into a destination golf course. This parcel, now known as the Waihe'e Coastal Dunes and Wetlands Refuge, is situated along Maui's northwest coast, approximately five miles outside of Maui's main town and seat of government, Wailuku (see fig. 1). The land trust acquired this property using local and federal funding, primarily because of its proximity to the major population centers on Maui, its historical and cultural significance, and the presence of important habitat features.

As part of the conditions of acquiring the permits needed for a proposed golf course development on the refuge in the 1990s, extensive archaeological work had been completed on the property prior to its purchase by MCLT. This research provided a strong foundation for understanding how the Hawaiian community had lived on the land for more than a thousand years. In total, researchers identified ninety-three archaeological sites and features, including three temple sites (*heiau*), an extensive village complex with evidence of over four hundred years of continual habitation, and the largest extant freshwater fishpond on Maui (*loko kalo i'a*), a portion of which makes up a twenty-seven-acre palustrine discharge wetlands complex. It soon became very clear that the Waihe'e refuge provided an important key for understanding how the indigenous Hawaiian community had lived interdependently with one another and with their environment for over a millennia. This clarity evolved into an awareness that Waihe'e told the local story of the life of the land over time, and that by reaching back into the past to understand the island worldview that shaped this community, MCLT might be able to reevaluate and transform its own work to see precontact Hawaiian sustainability in light of its own modern condition.

After meeting with key elders across Maui, MCLT was encouraged to meet with elders and opinion leaders in the Waihe'e community to better understand their connection to the land and their vision of how MCLT might contribute to its healing. Through this collaborative process, MCLT arrived at an

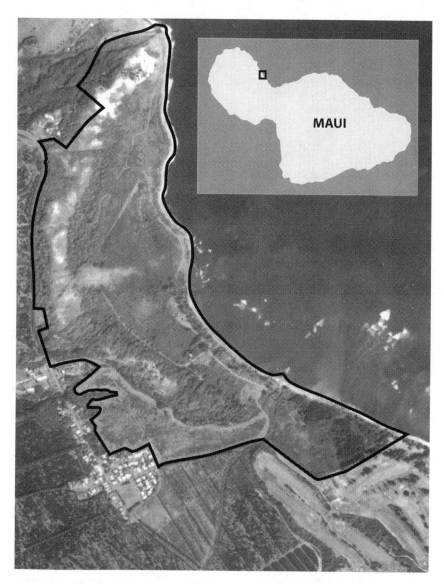

MAUI

FIGURE 1.1. The Hawaiian Islands Land Trust's 277-acre Waihe'e Coastal Dunes and Wetlands Refuge.

understanding that making the refuge a safe and welcoming place for all of Maui's residents constituted the most urgent and important work. The logic of the activity stemmed from the recognition that a community separated from its land could not cultivate the love of the land (*aloha 'āina*) that was needed to heal and care for it. Caring and healing the degraded ecosystems, particularly the wetlands with its abundance of endangered waterbirds, constituted a concrete strategy to foster a connection with the land, a key component of retrieving the island paradigm. As a result of the ecological restoration efforts, more than fifty-five acres of coastal strand and wetlands habitat have been restored, including much of its predisturbance structure, function, and floral and faunal composition.

A second objective that MCLT implemented in its efforts to make the refuge a safe and welcoming place involved protecting and interpreting the numerous historical and cultural sites on the refuge. In large part, this derived from having access to the historical and archaeological research that had been done on the refuge dating back to the early 1970s. All of this research pointed to the fact that Waihe'e, and the refuge in particular, was an important habitation site since very early in Hawaiian history. Several threads of evidence suggest this was true. First, several Hawaiian legends associate mythological individuals with the refuge and its immediate vicinity. For example, one myth ascribes the construction of the sand dunes that surround the refuge as the work of the goddess Haumea in her attempt to protect the sacred tree, Kalaukekahuli. Another myth suggests that Maui, the demigod, gathered the coconut fiber in Waihe'e that he later used to snare the sun on Haleakalā. While these myths are not conclusive evidence of Waihe'e's historical prominence, they do suggest that the Hawaiian community associated Waihe'e with its *mana*, or power.

A second strand of evidence for Waihe'e's importance comes from the early dates associated with its habitation. While the earliest definitive date, 941 AD, came from a study conducted in the late 1980s by an archaeologist from the Catholic University of America, a field survey in the early 1970s found fragmentary evidence of fishhooks with features characteristic of southern Polynesia. Because the Marquesas Islands are the presumed ancestral home of the early Hawaiian community, finding this artifact suggested that they had been made prior to the development of a distinctively Hawaiian style of fishhook.[10]

A third piece of evidence derives from an intimate awareness of the land, as seen from the island perspective, of the early Hawaiian community. Waihe'e possessed in abundance of all the features an early Hawaiian would have understood as critical to a thriving, abundant, and sustainable community. These included wetlands with sufficient seasonal moisture, where the community could cultivate taro and other food crops. Waiheʻe also had protected waters where nearshore gathering could take place. Perhaps most important, Waiheʻe also possesses several channels through the nearby reef where sailing and fishing canoes can safely traverse the reefs to access the nearby pelagic fishing grounds.

One of the most important activities on the refuge involves cultivating the awareness of the island worldview. Students frequently work alongside HILT staff conducting ecological restoration work, removing invasive species, improving endangered waterbird habitat, and assisting with the frequent monitoring on the refuge. By doing so, HILT staff hope to cultivate a sense of responsibility toward the land and to develop wisdom based on the fact that much can be learned by active practice. By retrieving this insight, HILT can facilitate the process of evaluating this awareness in light of the modern condition and transform this awareness into a modern context. In doing so, it points out the other critical strands of sustainability, specifically emphasizing communal interdependence and a deep connection to the land.

In order to ensure that students of all ages have access to the land, HILT developed a vigorous education program on the refuge. On average, more than two hundred students visit the refuge each month from schools all over the island, as well as occasional visits from off-island and out-of-state students. HILT employed a part-time education coordinator to ensure that the needs of both students and teachers are met while learning on the refuge.

Waiheʻe plays an important role in the practice of Hawaiian sustainability, primarily by representing a place where visitors to the refuge can better understand the way that the early Hawaiian community fostered a sense of interdependence with one another and with their land. The evidence of this interdependence remains a legacy of the land as it spans a period of more than one thousand years. Waiheʻe also stands out as a place where visitors can better understand, and participate in, the restoration of a degraded ecosystem. Finally, Waiheʻe represents a *wahi pana,* a sacred site, to retrieve, reevaluate, and transform the ancient Hawaiian society to make sustainability a meaningful concept for those living in the twenty-first century. This goal to connect people to the land is highlighted by Kikiloi (2010, 102), who points out that "the *ʻāina* sustains our identity and health by centering our attitudes, instincts, perceptions, values, and character within the context of our sacred environment."

CONCLUSION

The earliest Hawaiian communities brought with them an island worldview that emphasized an intimate and familial connection to the land based on interdependence. This worldview was predicated on understanding the intrinsic power (*mana*) of all phenomena and the need to protect this *mana* through the establishment of protective barriers, or *kapu*. For generations, as passed down through myths and legends, as well as proverbial sayings, Hawaiians understood that the life of the community was predicated on the health of the land. The arrival of foreigners in 1778 began an era of unprecedented change, through the introduction of both western arms and foreign diseases. The cumulative changes resulted in the dramatic loss of significant aspects of Hawaiian culture, the erosion of the island worldview in favor of a continental worldview, and ultimately the deterioration of Hawaiʻi's independence and sustainability.

Fortunately, enough vestiges of the original Hawaiian worldview remain to begin a coherent process of cultural revival that can simultaneously look to the past while attempting to answer some of the challenges of today. One of the most urgent problems lies in trying to recover a vestige of Hawaiʻi's sustainability. Numerous challenges lie ahead, but recovering a sense of interdependent sustainability remains a crucial element of this process. The connection to land remains a critical part of the process of fostering sustainability. As Kikiloi (2010, 75) points out, "At the core of this profound connection is the deep and enduring sentiment of *aloha ʻāina*, or love for the land. *Aloha ʻāina* represents our most basic and fundamental expression of the Hawaiian experience."

NOTES

1. The basic land divisions of Hawaiʻi, the *mokupuni*, or island, constituted the most fundamental and well-defined geographic territory. The *moku*, or district, made up the second most basic land division and frequently followed topographical features, such as ridge lines or the summits of mountains. Islands varied in their number of *moku* based on a number of factors, such as island size and political divisions.

2. Taro, or *kalo*, was the most basic Hawaiian staple, with the corm pounded into poi and the cooked leaves consumed as a green reminiscent of spinach. Hawaiian mythology postulates a sibling relationship between the Hawaiian people and taro. While taro was cultivated in both wet and dry contexts, the cultivation of wet taro in *loʻi* was practiced extensively throughout the archipelago.

3. Precontact Hawai'i was a highly stratified society, with the bulk of the population consisting of *maka'āinana,* who did the bulk of the work but were not obligated to remain on the land if they felt ill treated or felt the need to depart for any other reason. In this regard, *maka'āinana* in Hawai'i were unlike serfs of Western Europe.

4. The Hawaiians did alter their ecosystems for farming and other human-oriented activities.

5. Unlike most staples (wheat, corn, rice, etc.), taro does not store well for long periods. A consistent labor source is needed to quickly harvest, prepare, and consume taro in its preferred prepared form, poi (made from the pounded corm of the taro root).

6. In the English Standard Version, Genesis 1:26 states, "Let them [humans] have dominion over the fish of the sea, and over the birds of the heavens and over the livestock and over all the earth and over every creeping thing that creeps on the earth." Genesis 1:28: "And God blessed them. And God said to them, Be fruitful and multiply and fill the earth and subdue it and have dominion over the fish of the sea and over the birds of the heavens and over every living thing that moves on the earth."

7. The Hawaiian monarchy under Queen Lili'uokalani was illegally overthrown by a group of local businessmen, many of whom had dual American and Hawaiian citizenship, in January of 1893. In the interest of preventing bloodshed, the queen abdicated her throne, setting in motion the establishment of the Republic of Hawai'i, and later the Territory of Hawai'i, and in 1959 statehood.

8. Kamakau, Malo, and Ī'ī were Hawaiian historians who wrote from both academic and personal experience about the Hawaiian monarchy, traditional Hawaiian cultural practices, and the transition to western practices in Hawai'i. Their writings preserved much of Hawaiian culture and were critical to its preservation and eventual renaissance.

9. At the outset of World War II, through martial law, the US military claimed the entire island of Kaho'olawe for training purposes. In spite of the initial promises to return the island at the cessation of hostilities, the island remained under military jurisdiction until it was formally handed over to the State of Hawai'i in 2003.

10. This does not imply, however, that this implement was actually made in southern Polynesia. In all likelihood it was not. Rather, this fishhook was made using a conceptual design from southern Polynesia prior to the time when Hawaiians had developed their own unique style of fishhook design.

REFERENCES

Chun, M. N. 2009. "Ho'onohonoho: Traditional Ways of Cultural Management." Honolulu: Curriculum Research and Development Group, University of Hawai'i. Ka Wana Series, Book 9.

Cordy, R. 2000. *Exalted Sits the Chief: The Ancient History of Hawai'i Island.* Honolulu, HI: Mutual.

Daws, G. 1968. *Shoal of Time: A History of the Hawaiian Islands.* Honolulu: University of Hawai'i Press.

Ellis, W. 1963. *The Journal of William Ellis.* Honolulu, HI: Honolulu Advertising Company.

Handy, E. S. C., and E. G. Handy. 1991. *Native Planters in Old Hawai'i: Their Life, Lore, and Environment.* Honolulu, HI: Bishop Museum Press.

Kame'eleihiwa, L. 1992. *Native Lands and Foreign Desires.* Honolulu, HI: Bishop Museum Press.

Kikiloi, K. 2010. "Rebirth of an Archipelago: Sustaining a Hawaiian Cultural Identity for People and Homeland." *Hulili: Multidisciplinary Research on Hawaiian Well-Being* 6:73–115.

McGregor, D. P. 2007. *Nā Kua'āina: Living Hawaiian Culture.* Honolulu: University of Hawai'i Press.

Pukui, M. K. 1979. *Nānā i Ke Kumu: Look to the Source.* Vols. 1 and 2. Honolulu, HI: Queen Lili'uokalani Children's Center.

———. 1983. *'Ōlelo No'eau: Hawaiian Proverbs and Poetical Sayings.* Honolulu, HI: Bishop Museum Press.

Ritte, W. Jr., and R. Sawyer. 1978. *Nā Mana'o Aloha o Kaho'olawe: The Many Feelings of Love for Kaho'olawe.* Honolulu, HI: Aloha 'Āina o Nā Kūpuna.

Silva, N. 2004. *Aloha Betrayed: Native Hawaiian Resistance to American Colonialism.* Durham, NC: Duke University Press.

Thurston, L. C. 1934. *Pioneer Missionary in Kona, Hawai'i.* Ann Arbor, MI: S. C. Andrews.

Valeri, V. 1985. *Kingship and Sacrifice: Ritual and Society in Ancient Hawai'i.* Chicago: University of Chicago Press.

TWO

Food Security in Hawai'i

GEORGE KENT

According to the Food and Agriculture Organization of the United Nations, "food security exists when all people, at all times, have physical, social and economic access to sufficient, safe and nutritious food to meet their dietary needs for an active and healthy life" (2009, 8). Food insecurity can take many different forms. This essay explores three broad concerns for Hawai'i: overall food supply, disasters, and poverty. Each of these broad categories covers a variety of specific issues. For example, overall food supply is about food quantity and quality now and in the future, under various contingencies. It would include consideration of agriculture, processing, transport, infant feeding, nutrition-related health problems, genetically modified organisms, and related issues, under various long-term economic and climate scenarios. Disaster refers not only to tsunamis and earthquakes but also to economic collapse, terrorism, food supply crises, and other kinds of emergencies. Poverty refers to the difficulties in obtaining adequate food by various categories of low-income individuals, and also the status of the economy as a whole. Thus, food security must be recognized as multidimensional, raising a broad variety of concerns for which policies and planning are needed. In addition, Hawai'i must learn to differentiate food security, self-sufficiency, and resilience, which are related but are not the same.

Before contact, Hawai'i was self-sufficient in terms of food, by necessity, not by choice. There were periodic famines (*wi*), usually due to disruptive events such as epidemics and warfare (Schmitt 1970). After contact, Hawai'i became involved in exporting, sometimes with serious consequences: "Because the chiefs and commoners in large numbers went out cutting and carrying sandalwood, famine was experienced from Hawaii to Kauai. . . . The people were forced to eat herbs and fern trunks, because there was no food to be had. When Kamehameha saw that the country was in the grip of a severe famine, he ordered the chiefs and commoners not to devote all their time to cutting sandalwood" (Kuykendall, qtd. in Schmitt, 1970, 113). If the efforts used to harvest sandalwood had instead been devoted to harvesting or raising food, the famine could have been averted.

Hawai'i later devoted much of its land to producing and exporting food products. In the 1860s a new variety of rice was introduced, rapidly replacing taro. In 1862 Hawai'i exported more than 100,000 pounds of milled rice to California and more than 800,000 pounds of rice grown in paddies (Haraguchi 1987). Much of this production was by Chinese and Japanese immigrant laborers. Later, sugar and pineapple became the dominant export crops. Hawai'i was for a time a major exporter of food products, based on its sugar and pineapple production.

Much of Hawai'i's food production for local consumption has now been displaced by imported foods. This has raised alarm in many quarters. In 2012 the Hawai'i State House's self-sufficiency bill, HB2703 HD2, said Hawai'i is dangerously dependent on imported food:

> As the most geographically isolated state in the country, Hawaii imports approximately ninety-two percent of its food, according to the United States Department of Agriculture. Currently, Hawaii has a supply of fresh produce for no more than ten days. Ninety percent of the beef, sixty-seven percent of the fresh vegetables, sixty-five percent of the fresh fruits, and eighty percent of all milk purchased in the State are imported. The legislature further finds that Hawaii's reliance on out-of-state sources of food places residents directly at risk of food shortages in the event of natural disasters, economic disruption, and other external factors beyond the State's control. (Hawai'i State Legislature 2012)

Most analysts agree that Hawai'i currently imports 85 percent or more of its food from the US mainland and from other countries (Leung and Loke 2008; Page, Lionel, and Schewel 2007). Some analyses focus specifically on imports and local production of fresh fruits and vegetables (Lee and Bittenbender 2007; Southichack 2007).

The self-sufficiency bill (Hawai'i State Legislature 2012) raised the import-replacement argument: "The legislature further finds that each food product imported to Hawaii is a lost opportunity for local economic growth. The legislature notes that according to the University of Hawaii College of Tropical Agriculture and Human Resources, an increase in the production and sale of Hawaii-grown agricultural commodities would contribute to significant job creation. The research shows that replacing ten percent of current food imports will create a total of two thousand three hundred jobs." However, this analysis favors the producers' perspective and does not give sufficient attention to the consumers' perspective. Increasing purchases of locally produced foods would benefit local farmers, but it could also mean that consumers have to pay higher prices. The main reason Hawai'i imports much of its food is that it cannot produce the food as cheaply as it can import it.

Many people believe that the long-distance transportation of food to Hawai'i leads to high economic and environmental costs. However, there has been no broad study of these impacts. Ocean transport is relatively cheap, in terms of unit cost per mile. Its environmental impacts may be comparable to that of ground-based transport. Food prices in Hawai'i's stores may not relate as much to transport costs as to the fact that the retailers face less competition in Hawai'i than they do on the mainland and thus can charge higher prices. Higher real estate and utility costs also have to be considered.

Oceanic transport costs are not as high as many people assume. Hawai'i was once a major exporter of sugar and pineapple. Australia and New Zealand currently are major exporters of dairy products to much of the world. It is true that the Jones Act, which requires that shipping between US ports must be done on US-flagged ships (Maritime Trade Act of 1920 §27), increases transport costs, but quantitative estimates of the Jones Act's impact on food prices are not available.

Food security in Hawai'i is often understood in terms of possible interruptions to food imports, but there are other possible threats as well. For example, such things as local climate change, bee mites, and disruptions in water supply could threaten Hawai'i's agriculture. Local economic weaknesses of various kinds can lead to sharp reductions in local food production, as we have seen in dairy and meat production.

There are also dangers that can arise at the consumer end of the food system. Hawai'i has been fortunate so far in not having had any major food safety incidents, but there are safety risks. Hawai'i relies mainly on the federal government to ensure food safety.

Many people are concerned about the impacts of genetic modification of food products, especially the economic impacts at the primary production end and the health impacts at the consumption end. A large portion of Hawai'i's agricultural land is devoted to research on seeds for genetically modified products (Conrow 2009).

Expand Food Production?

While there is potential for increasing food production in local farms, there are huge challenges from competing uses of the land. First, some forces take land out of agriculture, such as encroaching housing developments and golf courses. Second, much of the agricultural land is used to produce crops other than food, such as seeds, biofuels, and ornamentals. Third, where there is food production, much of it is not basic food. Items such as coffee, macadamia nuts, and herbs will not be needed when food supplies are short. Fourth, some of the food that is produced is exported from the state. Fifth, it can be difficult for local farmers to compete with producers elsewhere who face lower land and labor costs.

In some cases the promotion of agriculture is mainly about protecting the livelihoods of small farmers, not about the products they deliver. For example, in the struggle to preserve the small farms in Kamilo Nui Valley in Hawai'i Kai, its defenders have not claimed this valley has been making an important contribution to the state's food supply. It is important as the basis for the livelihood of the farmers who work the land. Similarly, while the front page of the local newspaper may headline "Blight Threatens Basil" (Nakaso 2011), that evokes little concern about Hawai'i's basic food supply. The objectives of ensuring food security and protecting farmers' livelihoods are both important, but they should not be confused with one another.

Hawai'i's supply of land is limited, but its supply of ocean is not. However, food production in the ocean is difficult. Natural marine fisheries around the islands have never been highly productive because of the great depth of nearshore waters and the absence of nutrient upwelling associated with continental shelves. The reef fisheries have been severely depleted, so the great majority of fish consumption in Hawai'i is based on imports. There are attempts to revive traditional aquaculture methods, but they do not produce large volumes. Modern commercial aquaculture in Hawai'i has a checkered business record, with highly publicized ambitious start-ups often followed by quiet shutdowns. Some of the operations are owned and operated by businesses based outside Hawai'i and produce primarily for export, thus contributing little to the local food supply. There is evolving interest in aquaponics as an environmentally friendly method of combining aquaculture and farming.

Hawai'i's farm revenue (final crop output) set a record of $642 million in 1980 but it has decreased over the years, and was $633 million in 2012 (USDA 2014). With the steady decline of large-scale plantation agriculture, especially sugar and pineapple, average farm size declined to 149 acres in 2007. However, it then rose to 161 acres in 2012, primarily because of the increased share of land used for seed production. About 92 percent of Hawai'i's farms are less than 100 acres. In 2012, 58.4 percent of the farms had annual sales under $10,000. The net farm income in 2012 was estimated at $329,964,000 (USDA 2014), which means that in Hawai'i 7,000 farms had an average income of $47,138. The income levels for small farms were much lower than this average (Gomes 2011; USDA 2014). Much of the farm revenue is for nonfood products such as seeds and ornamentals and for exports. Hawai'i's farm revenue attributable to food consumed within the state is about $400 million per year. Hawai'i's total food imports are roughly $2 billion per year. On this basis, Hawai'i farms produce roughly 20 percent of the state's food supply, in terms of monetary value. Probably about 80 percent of the imports are from the US mainland. A substantial share of the food produced and consumed in Hawai'i goes to military families and tourists. Perhaps that share should be excluded from calculations about the degree to which local agriculture contributes to local food self-sufficiency.

Some local food production operations are owned by outsiders. Their products may be sold and consumed locally, but if the profits go elsewhere, and control of these operations also is based elsewhere, it is not clear that these operations really contribute to local self-sufficiency. Many of Hawai'i's agriculture workers have been immigrants, and that pattern is likely to continue in the future. The significance of this for Hawai'i's self-sufficiency should be given some thought.

As in other high-income places, most of Hawai'i's food money goes to processors, not farmers, and most of this goes to processors outside Hawai'i. Only about 7,300 people work in food processing in Hawai'i (Yonan 2011). There are efforts to expand local food processing (Hawai'i Food Manufacturers Association 2011). There are opportunities to expand the food processing sector, but the potential is limited because Hawai'i's processors must work with high costs and small volumes and compete with large-scale processors based elsewhere.

While it is true that Hawai'i's physical environment could support production of a wide variety of food items, the high land and labor costs make increasing production difficult. Moreover, Hawai'i could not produce the wide variety of products it now imports. Full self-sufficiency would be impractical, and getting close to full self-sufficiency would require radical changes in diet and lifestyle. Some people would welcome those changes, and others would not.

Many people outside Hawai'i call for localizing food production everywhere, but the arguments for localization may be overstated (Dean 2007; Singer and Mason 2007; Desrochers and Shimizu 2008; DeWeert 2009; McWilliams 2007; Roberts 2009). In Hawai'i, pushing food self-sufficiency too far or in the wrong way could increase costs to consumers, and it could reduce local food security by creating overdependence on one source. If it is not managed well, it could lead to the depletion of local resources. Increasing self-sufficiency could be advantageous to certain groups in the state, such as farmers, while being disadvantageous to others, such as the nonfarming poor.

There is a great deal of enthusiasm for increasing local food self-sufficiency, in the state government and in the community (see, for example, http://hawaiihomegrown.net). However, there is a need for discussion about how far and how fast it should go. The degree of self-sufficiency is not something that should be maximized. It should be optimized, taking a broad variety of issues into consideration.

It would be good for Hawai'i to have the *capacity* to be food self-sufficient in case it was suddenly isolated from the rest of the world. But if Hawai'i pushes for *actual* self-sufficiency long before it is needed, its people would forgo the benefits that come with trade. It would be a bit like moving the family into the basement now because a storm is likely to come in the next few years. Preparing is one thing; doing it is something else.

Resiliency

According to the 2012 food self-sufficiency bill, "increasing local production will ensure that Hawai'i's food sources will be more resilient to global supply disruptions, better able to cope with increasing global demand and shortages of commodities such as oil, and better prepared to deal with potential global food scarcities" (Hawai'i State Legislature 2012). That needs to be explained, and its limits should be appreciated.

Worldwatch defines "resilience" as:

> the ability of natural or human systems to survive in the face of great change. . . . To be resilient, a system must be able to adapt to changing circumstances and develop new ways to thrive. In ecological terms, resilience has been used to describe the ability of natural systems to return to equilibrium after adapting to changes. In climate change, resilience can also convey the capacity and ability of society to make necessary adaptations to a changing world—and not necessarily structures that will carry forward the status quo. In this perspective, resilience affords an opportunity to make systemic changes during adaptation, such as addressing social inequalities. (2009, 203)

On this basis, resilience in a food system would mean being able to choose from a variety of alternative food sources and being ready to jump from one to another in an agile way with changing conditions. Resiliency is different from self-sufficiency. Food security, in the sense of ensuring access to food under all conditions, comes mainly from resiliency, not self-sufficiency.

Hawai'i should have a variety of food sources available so that if one fails or weakens, it would be possible to shift to other sources. Hawai'i already does that on a regular basis for fresh produce, with wholesalers jumping around to different sources opportunistically. Increasing Hawai'i's capacity to produce its own food would increase its resilience to the extent that it added another source of food. However, if it displaced other sources, the result could be decreased resilience.

Hawai'i should not pursue food self-sufficiency to the extent that it allows its contacts with other sources of food to wither away. Just as Hawai'i should not be overly dependent on imports, it should not be overly dependent on its own production.

DISASTERS

Hawai'i has had a long run of good fortune, but it is not immune from disasters. Given its huge dependence on imports, the state has to be especially concerned about possible disruptions in transport to the islands. In 1949, when it was still a territory of the United States, Hawai'i suffered through a shipping strike, and questions arose about what the US government would do to help (*Time* 1949). Hawai'i is now a US state, but it is still not clear what help the US government would offer if Hawai'i, and possibly the US government itself, encountered some sort of extreme situation.

"Disaster" is defined by the UN International Strategy for Disaster Reduction as "a serious disruption of the functioning of a community or a society causing widespread human, material, economic or environmental losses which exceed the ability of the affected community or society to cope using its own resources" (UN Office for Disaster Risk Reduction 2006). Resilience can be understood as the capacity to make adaptations to the existing food system in response to changes in the physical or economic environment. In dealing with slow and permanent changes, it is about creating a new kind of "normal." In disaster planning, however, the concern is to find ways to prepare for quick changes, especially unanticipated quick changes. Usually disaster planning is based on the hope that the impact will be of short duration and that it will be possible to return to the same basic food system that existed before the disaster. In extreme disasters that system may need to be reconfigured with great urgency.

Hawai'i has not yet had major problems with its overall food supply, but there is a need for concern because Hawai'i imports so much of its food. Disruptions to that delivery system could be disastrous, especially if the disruption is sudden and Hawai'i is unprepared.

Production

Preserving and expanding Hawai'i's farm acreage alone would not be enough to ensure future food security. If there were a sudden cutoff in imported food, we would need a rapid switchover from production of nonfoods and nutritionally unimportant foods (e.g., coffee, macadamia nuts, herbs) to basic foods to ensure that everyone is well nourished. Plans should be made well in advance to facilitate such a conversion if and when it should become necessary. Historical wartime mobilizations suggest the possibilities for rapidly increasing local production of basic food.

In extreme emergencies, national and local governments might not be able to cope. Thus, some people focus on household and local food production, taking measures that are independent of government initiatives. For many people this is an ideological issue, based on the premise that even in good times, families and local communities ought to depend mainly on foods that they themselves produce. Some survivalists take this to an extreme, and many others do these things in a more limited way.

Storing Food

To deal with emergencies, it is important to work not only on food production but also on food storage, at the state level, in communities, and in households. Household food storage is increasingly important because the major food sellers no longer maintain large warehouses. The just-in-time delivery system has sharply reduced the merchants' need for warehouses, so if shipping to the state were to be suddenly cut off, the supply of food would last no more than a few days. Many people store nonperishables and water supplies in their basements or closets. Many groceries now sell specially designed emergency food supplies to be stored at home.

Historically, many places have identified particular famine foods. Sweet potatoes are especially good for this purpose and could be grown in many places that are otherwise unused, such as forests and meadows (Kristof 2010). In Hawai'i, 'ulu (breadfruit) played an important role in protection against disasters. Hawai'i should prepare for many different kinds of contingencies. The state could be deeply affected by disasters locally, as well as by disasters elsewhere if they interrupt the flow of food to Hawai'i. For example, if bees stopped pollinating in Hawai'i, its agriculture system could weaken or even collapse. It would then have to import more food. If the bees quit working in some places outside Hawai'i, it could import from other places. If the bees quit everywhere, everyone would be in trouble.

Hawai'i should be concerned not only about actual shortages but also about anticipated shortages. If rumors build up about possible shipping interruptions, there could be a run on food stores. There is no evident governmental plan for dealing with hoarding before, during, or after disaster events.

At the global level, speculation in food commodities can be viewed as another form of hoarding, one that could result in increased food insecurity for many people. The great global land grab, in which rich countries are gaining control over poor countries' agricultural resources to ensure their own future food security, is another form of hoarding at the global level (Center for Human Rights and Global Justice 2010). Hawai'i is not immune from such forces. To illustrate, if Hawaii's regular sources of rice suddenly diverted their production to other buyers, Hawai'i would be in serious trouble. The state is not likely to restore Waikiki to rice production. In prolonged emergencies there might be a need for food rationing of some sort. In extreme situations there might be a need for martial law, as there was in the 1940s (Bennett 1942).

As indicated above, the United Nations defines disasters as situations that are beyond the coping abilities of any particular place. This means that in disaster planning we must go beyond strengthening the capacities of individuals, families, and communities. There is a need to work out systems for assistance among different places. This could mean systems of support from one *ahupua'a,* or land division, to another, one island to another, or the entire state of Hawai'i to the United States, other nations, or the global community as a whole.

Despite Hawai'i's vulnerabilities, these relations have not been worked out with the clarity and foresight that is needed. If Hawai'i had a big problem with its food supply, it might be able to get help from the outside, but there are huge uncertainties. Where would the aid come from, on what terms? Some people might assume that the US government would come to Hawai'i's assistance under various contingencies, but we don't know for sure. How long would the US government help? In what ways? Are there commitments in writing? What if the entire United States faces an emergency and becomes unable to come to Hawaii's assistance? Where else could Hawai'i direct its appeals for help?

Ideas on how to approach these issues are suggested by the Model Intrastate Mutual Aid Legislation, available through the Hawai'i State Civil Defense website (http://www.scd.hawaii.gov/nims.html). Much work remains to be done on this. The State Civil Defense system focuses on hazards such as tsunamis and earthquakes but does not give attention to such things as shipping interruptions or disruptions in the state's agriculture.

Attention should be given to the food-related dimensions of disasters such as tsunamis and earthquakes. In all disasters, food-related problems begin to show up as soon as the warnings begin, with runs on stores. Any sort of prolonged disaster would raise serious concerns about food. There are also possibilities for food-centered disasters that have nothing to do with tsunamis and earthquakes.

Plans should be made for dealing with food crises regardless of the cause of the disruption. The benefits would far outweigh the costs. However, there is currently no clear mandate for any agency of state government to undertake this work.

Poverty-based food insecurity occurs in high-income as well as low-income countries. A great deal could be learned from the way it is addressed in other high-income countries (e.g., Sydney Food 2007). In many countries the problem of food security for the poor is given little attention, but it occurs in some degree everywhere.

The US federal government has been undertaking regular studies of food insecurity, focusing on the type that is associated with poverty (USDA 2010). Adapting the USDA's methods, the Hawai'i Department of Health in 2001 went into the issues much more deeply. It concluded that "food insecurity was prevalent in Hawai'i: one in six (16.4%) households and 1 in 5 (19.2%) individuals experienced either being at risk of hunger or experiencing hunger in 1999–2000. The poor, children, single adult households, and Pacific Islanders were particularly vulnerable" (Hawai'i Department of Health 2001).

The map in figure 2.1 indicates the geographical distribution of food insecurity. In Waimanalo, Wai'anae, Puna, Ka'a'wa, and Moloka'i, in 2001 more than 30 percent of the people lived in households that were not sure how they would get their food. Because of the high cost of living, many people who are not officially poor suffer from food insecurity (Hawai'i Department of Health 2001). The official poverty rate in Hawai'i hovers around 10 percent. Among the different ethnic groups, Native Hawaiians have the lowest average family income (Kana'iaupuni, Malone, and Ishibashi 2005). The impact is clear in the distribution of food insecurity in Hawai'i.

Another study on poverty-related food insecurity in Hawai'i (Giles, Shireen, and Derrickson 2002) used the USDA framework for assessing food insecurity, but went further by sketching out a proposed Community Food Security Plan. It emphasized the need for action by the state legislature and described several bills that were submitted to the legislature, but they were not passed.

According to USDA estimates, averaging for the years 2007–2009, 11.4 percent of Hawaii's households had low or very low food security, compared with 13.5 percent for the United States as a whole; 3.9 percent of households in Hawai'i had very low food security, compared with 5.2 percent in the United States as a whole (USDA 2010). Thus, Hawai'i has done relatively well. Nevertheless, poverty-based food insecurity is a persistent issue in the state, and as indicated above, the prevalence is higher among particular groups.

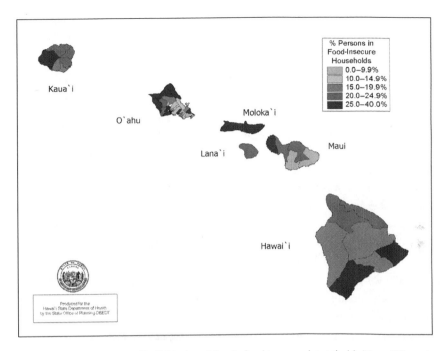

FIGURE 2.1. Percentage of individuals residing in food-insecure households (Hawaiʻi Department of Health 2001).

The USDA had to take into account the extraordinarily high price of food in only two states. "For residents in Alaska and Hawaii, the Thrifty Food Plan costs were adjusted upward by 19 percent and 63 percent, respectively, to reflect the higher cost of the Thrifty Food Plan in those States" (Nord 2010). Higher food prices mean greater food insecurity for much of the state's population, not just the very poor.

The Hawaiʻi Foodbank describes itself as "the only nonprofit 501 (c)3 agency in the state of Hawaii that collects, warehouses and distributes mass quantities of both perishable and non-perishable food to 250 member agencies as well as food banks on the Big Island, Maui and Kauai" (2010). In one year the Hawaiʻi Foodbank, through its cooperating agencies, served 183,500 different people in the state, including more than 55,000 children and more than 11,000 seniors. According to the Foodbank (2010):

- 79 percent of client households served are food insecure, meaning they do not always know where they will find their next meal.

- 43 percent of these client households experience food insecurity with hunger, meaning they are sometimes completely without a source of food.
- 83 percent of client households with children served are also food insecure.
- Of the 183,500 people the Hawai'i Foodbank network serves:

  - 79 percent of households have incomes below the federal poverty line.
  - The average monthly income for client households is $850.
  - 42 percent of households have one or more adults who are working.

Each year the Foodbank organizes large-scale campaigns to collect nonperishable food products from many different donors. It then provides food at little or no cost to such agencies as Aloha Harvest, the Institute for Human Services, Salvation Army, Waikiki Health Center, River of Life Mission, Kau Kau Wagon, Harbor House, and many church pantries so that they can respond to food insecurity and related problems. The programs that hand out food to the needy do a good job of tiding people over, but many unmet needs remain.

The Foodbank periodically raises the alarm about widespread hunger in the state when it conducts its food collection drives, but historically the state government has said very little about the issue. This may leave people uncertain as to whether it is really a serious problem in Hawai'i.

The state government administers the hundreds of millions of dollars that come into the state each year for federally funded nutrition programs such as school meals, the Supplemental Nutrition Assistance Program (SNAP; formerly Food Stamps), and the Special Supplemental Nutrition Program for Women, Infants, and Children, commonly known as WIC. However, apart from that, the state has not addressed the problem of poverty-based food insecurity. It has taken little notice of the data on food insecurity in Hawai'i that are provided each year by the US Department of Agriculture. The Hawai'i Department of Health used to include food-security questions in its annual health survey, but it no longer does, and it has not updated its 2001 study *Hunger and Food Insecurity in Hawai'i*.

Poverty-based food insecurity in Hawai'i is not high by global standards, but it exists, and it contradicts the image the state tries to portray of the quality of life in the islands. Hawai'i does not provide a strong safety net for all of its people. State officials may feel that the coverage by federally funded programs such as SNAP and WIC, together with the work of the nongovernmental organizations, is enough to meet the needs. However, there is a need to determine whether that is so and to consider what should be done for those who fall through the cracks.

The state's inattention to the poverty-based food security issue may be partly due to the fear that dealing with it could be costly. However, many helpful things could be done at low cost. The state could do more to pursue federal grants for community nutrition, such as those available through the US Department of Agriculture. The state's modest support for the local nongovernmental groups working on the issue, such as the Hawai'i Foodbank, seems to have yielded considerable benefits for a very small investment.

Many people who are eligible for SNAP and WIC do not take advantage of their services. The state, working together with interested nongovernmental organizations, could encourage more eligible people to apply. Hawai'i's legislature could learn from the ways in which other states invest a small amount of resources to help their people take full advantage of federal programs (e.g., Illinois General Assembly 2004).

There are other opportunities to draw in benefits for the poor, even if they are not specifically food oriented. For example, it has been estimated that as many as 34,000 taxpayers in Hawai'i may not be applying for the Earned Income Tax Credit to which they are entitled. The Family and Individual Self-Sufficiency Program at the Hawai'i Alliance for Community-Based Economic Development offers help along these lines, but more could be done (Tanna 2010).

When the state is going through a difficult time economically and cutting back on public services, it should be giving more attention to the food security issue, not less. This does not necessarily mean that the state has to provide more direct services. It should monitor the issue and call for help where it is needed.

The challenge is not to feed people but to see to it that they live under conditions in which they can provide for themselves. Dignity comes from providing for yourself and your family, not from standing in a soup-kitchen line. All able-bodied people should have decent opportunities to take care of themselves. Regardless of whether we draw on federal resources or charitable giving or local farmers' markets, the state government should take the responsibility to ensure that no one in the state remains food insecure.

It is not only poverty-related food security that the state has ignored. Until recently, the state government in Hawaiʻi has given little attention to the security of the overall food supply, and it has not done disaster planning related to possible food crises. Whether we are concerned with sudden-onset disasters or the threats to the food supply that come with slow climate change and increasing energy costs, there are compelling reasons for serious planning. Hawaiʻi's food system should be designed to be as resilient as possible so that it is prepared to deal with all sorts of changes in conditions. Ensuring good nutrition for all segments of the population under all conditions is a challenge that requires sustained attention.

In 2002 and early 2003, with prodding from interested citizens, the state legislature asked the Office of Planning in the Hawaiʻi Department of Business, Economic Development, and Tourism to convene a Food Security Task Force, to examine the best ways to ensure food security for Hawaiʻi's people. As a result of that group's work, in 2003 the legislature considered bills to create a permanent state Food Security Council. As stated in the conclusion of the task force's report:

> Hawaiʻi has no State, county or local food policy council to coordinate or oversee food security activities. Without State policies, objectives, or goals to guide State actions, no organization can effectively coordinate assistance programs, conduct ongoing monitoring, or spearhead integrated planning programs. With an adequate State match (funds, personnel), on an on-going basis, the State could leverage available federal dollars for food security coordination, food stamp outreach and education, and farmers markets initiatives, which can then be used to enhance food security and put food dollars into the pockets of the needy, local farmers and food retailers thereby spurring our economy from the ground up. (2003, 14)

The idea was that the council, including both government officials and private citizens, would envision a food-secure Hawaiʻi and then try to figure out how to get there. The council would bring together all concerned parties to formulate a coherent strategy for identifying and addressing the issues. However, the legislature did not approve the proposal.

Given the persistent need to strengthen Hawaiʻi's food security, interested individuals and organizations gathered together in November 2010 to establish a nongovernmental, community-based Hawaii Food Policy Council (HFPC; Lukens 2010). As explained at its website (http://www.hawaiifood policycouncil.org/) and its Facebook page (http://www.facebook.com /HawaiiFPC) the HFPC's primary role is to provide a forum for exploring the major food security issues confronting the state. Without the engagement of the state government, the HFPC's capacities would be very limited. There is a need for an interagency unit in the state government that would have primary responsibility for ensuring food security for all parts of the state's population under all conditions. This unit could work together with the community-based HFPC and serve as a major channel through which the government would hear the concerns of the people. Hawai'i's government and people need to act together to strengthen the local food system, and address the full range of food security issues that confront it.

REFERENCES

Bennett, M. K. 1942. "Hawaii's Food Situation." *Far Eastern Survey* 11 (16): 173–175.

Center for Human Rights and Global Justice. 2010. *Foreign Land Deals and Human Rights: Case Studies on Agricultural and Biofuel Investment.* New York: Center for Human Rights and Global Justice, New York University School of Law. http://www.business-human rights.org/Links/Repository/1003276.

Conrow, J. 2009. "A Seed of Doubt." *Honolulu Weekly* 19 (April 8–14): 6–7. http://hono luluweekly.com/cover/2009/04/a-seed-of-doubt/.

Dean, A. 2007. *Local Produce vs. Global Trade.* New York: Carnegie Council: Policy Innovations, October 25. http://www.policyinnovations.org/ideas/briefings/data/local _global.

Desrochers, P., and H. Shimizu. 2008. *Yes, We Have No Bananas: A Critique of the "Food Miles" Perspective.* Arlington, VA: George Mason University. http://mercatus.org /publication/yes-we-have-no-bananas-critique-food-miles-perspective.

DeWeert, S. 2009. "Is Local Food Better?" *Worldwatch,* May–June. http://www.worldwatch .org/node/6064?emc=el&m=227941&l=4&v=67949a0ab6.

FAO 2009. *The State of Food Insecurity in the World.* Rome: Food and Agriculture Organization of the United Nations.

Food Security Task Force. 2003. *A Report to the Legislature on SCR 75, SD1, HD1, 2002.* Honolulu: Office of Planning, State of Hawai'i. http://hawaii.gov/dbedt/op/fstfr_2003.pdf.

Giles, C., Z. Shireen, and J. P. Derrickson. 2002. *Hawai'i Community Food Security Needs Assessment.* Washington, DC: Congressional Hunger Center and Full Plate.

Gomes, A. 2011. "State Farm Revenue Near Peak." *Honolulu Star-Advertiser,* January 16, D1. http://www.staradvertiser.com/business/20110116_State_farm_revenue_near _peak.html.

Haraguchi, K. 1987. *Rice in Hawaii: A Guide to Historical Resources.* Honolulu: Humanities Program of the State Foundation on Culture the Arts and Hawaiian Historical Society.

Hawai'i Department of Health. 2001. *Hunger and Food Insecurity in Hawai'i: Baseline Estimates.* Honolulu: Hawai'i Department of Health.

Hawai'i Foodbank. 2010. *Hunger Facts.* Honolulu: Hawai'i Foodbank.

Hawai'i Food Manufacturers Association. 2011. Home page. http://www.foodsofhawaii.com.

Hawai'i State Legislature. 2012. "Hawai'i State Legislature: Bill Status and Documents." http://www.capitol.hawaii.gov/.

Illinois General Assembly. 2004. *Public Aid: (305 ILCS 42/) Nutrition Outreach and Public Education Act.* http://www.ilga.gov/legislation/ilcs/ilcs3.asp?ActID=2476&ChapterID=28.

Kana'iaupuni, S. M., N. J. Malone, and K. Ishibashi. 2005. *Income and Poverty among Native Hawaiians: Summary of Ka Huaka'i Findings.* Honolulu, HI: Kamehameha Schools.

Kristof, N. D. 2010. "Bless the Orange Sweet Potato." *New York Times,* November 24. http://www.nytimes.com/2010/11/25/opinion/25kristof.html?_r=1&emc=tnt&tntemail1=y.

Lee, C. N., and H. C. Skip Bittenbender. 2007. "Agriculture." In *Hawai'i 2050: Building a Shared Future: Issue Book.* http://hawaii2050.org/images/uploads/HI2050_web5.pdf.

Leung, P., and M. Loke. 2008. *Economic Impacts of Increasing Hawaii's Food Self-Sufficiency.* Honolulu: University of Hawai'i. http://hdoa.hawaii.gov/add/files/2012/12/Food SSReport.pdf.

Lukens, A. 2010. "Democratizing Food: Hawai'i's Future Food Systems." *Honolulu Weekly,* November 24. http://honoluluweekly.com/feature/2010/11/democratizing-food/.

McWilliams, J. E. 2007. "Food That Travels Well." *New York Times,* August 6. http://www.nytimes.com/2007/08/06/opinion/06mcwilliams.html?_r=1&oref=slogin.

Nakaso, D. 2011. "Basil Bane Putting Bit on Business." *Honolulu Star-Advertiser,* February 3. http://www.staradvertiser.com/news/20110203_Basil_bane_putting_bite_on _business.html.

Page, C., B. Lionel, and L. Schewel. 2007. *Island of Hawai'i Whole System Project: Phase 1 Report.* Snowmass, CO: Rocky Mountain Institute. http://www.kohalacenter.org/pdf /hi_wsp_2.pdf.

Roberts, P. 2009. "Spoiled." *Mother Jones* 24 (2): 28–36.

Schmitt, R. C. 1970. "Famine Mortality in Hawai'i." *Journal of Pacific History* 5: 109–115.

Singer, P., and J. Mason. 2007. *The Way We Eat: The Ethics of What We Eat and How to Make Better Choices.* Emmaus, PA: Rodale Press.

Southichack, M. 2007. *Inshipment Trend and Its Implications on Hawaii's Food Security.* Honolulu: Hawaii Department of Agriculture. http://www.kohalacenter.org/pdf/HDOA _hawaii_food_security.pdf.

Sydney Food. 2007. *Understanding Food Insecurity: Why Families Go Hungry in an Affluent Society.* Sydney, Australia: Sydney Food Fairness Alliance and Food Fairness Illawarra. http://sydneyfoodfairness.org.au/wp-content/uploads/2009/07/SFFA_insecurity _v1_aug07www.pdf

Tanna, W. M. 2010. "Earned Income Credit Is Bright Start of Tax Policy." *Honolulu Star-Advertiser,* January 27. http://www.staradvertiser.com/editorials/20110127_earned _income_credit_is_bright_star_of_tax_policy.html.

Time. 1949. "Labor: Who Gives a Damn?" July 4. http://content.time.com/time/magazine/article/0,9171,888526,00.html.

UN Food and Agriculture Organization of the United Nations. 2009. *The State of Food Insecurity in the World*. Rome: FAO. http://www.fao.org/docrep/012/i0876e/i0876e00.htm.

UN Office for Disaster Risk Reduction. 2006. *Terminology: Basic Terms of Disaster Risk Reduction*. Geneva, Switzerland: United Nations. http://www.unisdr.org/eng/library/lib-terminology-eng-p.htm.

USDA (US Department of Agriculture). 2010. "Table: Household Food Security Status by State." Washington, DC: US Department of Agriculture.

———. 2014. *State Fact Sheets: Hawaii*. Washington, DC: US Department of Agriculture. http://www.ers.usda.gov/data-products/state-fact-sheets/state-data.aspx?StateFIPS=15&StateName=Hawaii#.VBtun-e7kbI.

Worldwatch. 2009. "Climate Change Reference Guide and Glossary." In *State of the World 2009: Into a Warming World*. Washington, DC: Worldwatch Institute.

Yonan, A. Jr. 2011. "Manufacturing Slump Slows." *Honolulu Star-Advertiser*, February 22. http://www.staradvertiser.com/business/businessnews/20110222_Manufacturing_slump_slows.html.

THREE

Searching for Sustainable Agriculture in Hawai'i

PENNY LEVIN

Aia i ka mole ke ola; e 'ike pono i ke au nui me ke au 'iki
There in the foundation is life; know well the big currents
and the little currents

Sustainable agriculture is a common turn of phrase. What does it really mean, particularly in an island context? This chapter explores how the concept of sustainable agriculture is being defined and redefined or, one could argue, restored in Hawai'i, through the traditions and resurgence of *kalo* (taro; *Colocasia esculenta*) farming, the first, oldest, and culturally most significant food crop in the state.

The *mo'okū'auhau* (genealogy) of agriculture in Hawai'i is one that can be described as beginning with the sharing of *hā* (breath) between farmer and a living landscape and arriving in the present to a place largely out of touch not just with its legendary origins, but with the microelements, connective tissue, and pulses requisite of a sustainable system. Understanding what has been lost—the small and the large transitions that have moved agriculture in Hawai'i away from its roots—over the last three hundred years also offers a *mole* (pronounced mo-leh; a foundation) from which to regenerate good practice.

A BRIEF HISTORY OF THE UNSUSTAINING OF AGRICULTURE IN HAWAI'I

Archaeological evidence has suggested initial habitation in the Hawaiian Islands around 300–500 AD, roughly 1,700 years ago (Kirch 1985; Athens 1997).

Kalo was elder brother to 1st Hawaiian.

Kirch (2011) has since revised his assessment to 1000 AD based on new records in the Pacific and more refined dating technologies. The earliest foods grown specifically for human consumption were the "canoe plants," an estimated twenty-five useful species that traveled with the Polynesians who became the first Hawaiians to these islands (White 1994; Abbott 1992; Sohmer and Gustafson 1987; Kirch 1985).[1] All were important, but among them, *kalo,* *'uala* (sweet potato; *Ipomoea batatas*), and *mai'a* (banana; *Musa* sp.) were the foundation crops upon which health and survival were built.[2] It was *kalo* that held the most sacred place in the relationship between man, food, and the land (Handy, Handy, and Pukui 1991, 3–4).

As told in oral tradition and written genealogies, Hāloanakalaukapalili, the long-stemmed trembling leaf, the first *kalo* plant, was elder brother to Hāloa, the first Hawaiian man (Beckwith, 1970; Liliuokalani of Hawai'i 1897). The corm (steamed and eaten, or pounded into *pa'i 'ai* or poi),[3] stalks and leaves (steamed or boiled, as *lū'au*), and flowers (steamed) of the *kalo* provided food and medicine, and man provided the labor to care for the plant. That was the reciprocal agreement between elder and younger siblings.

As Hawaiians multiplied, so too did the *kalo,* into hundreds of unique varieties and across thousands of acres. Some varieties were designated *kinolau,* or body forms, of the Hawaiian gods and became appropriate offerings woven into the complex ritual and cosmology of the Hawaiian world. Keen observers of the cycles and behaviors of the land, Hawaiians developed unique *lo'i kalo* (irrigated, wetland taro field) and *'āina malo'o* or *māla 'ai* (a dry, rainfed field) systems and practices fit to a diverse set of growing conditions and topographies that in their height, supported a population comparable to the present day in the islands.[4]

Sugar, pineapple, ranching, and invasive plant species, along with the short-lived sandalwood trade, have each taken their toll in Hawai'i since their introduction beginning in the late eighteenth century. Each created radical changes in food availability, natural resource cycles and conditions, and agricultural productivity, the most dramatic being the permanent movement of water out of streams and watersheds in the last 150 years, which disconnected the flow of nutrients and cool, fresh water between the mountains and their receiving reefs. This ran counter to the traditional Hawaiian wisdom of returning water to its source within an *ahupua'a* (a traditional land division that often embraced a stream or gulch and its surrounding ridges from the uplands into the coastal waters at its base), a practice that ensured that a diverse, land- and ocean-based food system remained intact.[5]

The history of sociocultural impacts that went hand in hand with the western mindset of the whalers, businessmen, and missionaries who came to these islands cannot be addressed in so small a space as this chapter. It is important to note, however, that with the scale of physical changes in the landscape and subsequent shifts in agrarian practice led by export-driven agricultural models, came a rending of the sacred and practical relationships between a people and the very breath of life, elements that played pivotal roles in the development of sustainable agriculture systems in pre-contact Hawai'i.[6]

Captain Cook's arrival in 1778 is the seminal date by which historians and researchers mark the turning point and decline of the Hawaiian people, their culture, and practices. It also marks the beginnings of written records of the flora, fauna, culture, and agriculture of the Hawaiian Islands.[7] Prior to that time, we must rely on recorded oral tradition, botanical and archaeological evidence to understand changes in the islands' endemic landscape that influenced the course of traditional agriculture.

The precontact modified landscape described by Handy, Handy, and Pukui (1991, 224) is confirmed by more recent paleoecological evidence. Studies indicate that changes in native plant species composition between pre- and post-Hawaiian settlement may have been as high as 80–100 percent in the lowlands and "extensively altered by human inhabitants" up to 1,500 feet in elevation by 1500–1600 AD (Burney and Kikuchi 2006; Athens 1997; Kirch 1982). Cox (1991, 2) noted that "the coastal Hawaii that Cook, Vancouver and other early visitors saw was a cultural landscape shaped by human hands, not a natural one."

Growing populations, the demands of an increasingly complex hierarchy of ali'i (chiefs) and the numerous wars between chiefs impacted ecosystem conditions in significant ways, which in turn resulted in cultivation intensity, feast and famine, and the decline of some of the largest dryland food production systems prior to Cook's arrival (Altonn 2004; Vitousek et al. 2004; Allen 2001; Kirch and Hunt 1997; Kirch 1994). Archaeological dates associated with a variety of agricultural sites, coupled with archival, ethnographic, and Hawaiian newspaper records, suggest increasingly detailed and complex food production practices that included landscape-level and microtopographic management strategies for forests, water, soil, planting stock, and food systems or "regulated resource use" within each ahupua'a (Ladefoged et al. 2009; Kamakau 1992; Merlin and VanRavenswaay 1990; Kelly 1989), as a necessary response to declining environmental health.

The histories and traditions of the *ali'i* are more consistently recorded than the details of life of the *maka'āinana* (people who tended the land; commoners); hence, what occurred in the fields is far less well known from this era. Of the hundreds of *kalo* names recorded, there is no way to know for certain which ones were present before written record and which names may have evolved after 1800, or the exact number of unique cultivars. However, a second likely response to landscape and population changes would have been peaks and ebbs in crop cultivar diversity. The intensification of *kapu* systems (prohibited or privileged activities, things, or places) and *ahupua'a* boundaries restricted and defined people's movements and trade patterns. While the *ali'i* may have travelled extensively, Malo (1898) suggests that *maka'āinana* remained primarily within a district. In the natural environment, isolation induces uniqueness, less sharing of crop varieties across *ahupua'a* boundaries, and hence more local specialization (this mimics the behavior of Hawai'i's endemic species).[8] Conversely, the impacts of continual wars may have resulted in a need for fast-growing cultivars to feed warring parties on the go and a loss of some varieties as fields were abandoned or destroyed to prevent a district from feeding itself.[9] People fled into the mountains or to other islands ahead of battles and carried *kalo* with them.[10] With the consolidation of the islands, ruling chiefs governed larger and larger tracts of land and hence influenced more villages and larger populations. While it is difficult to find concrete evidence, it makes sense that the *kalo* varieties preferred by these *ali'i* would have been distributed throughout their lands (particularly in *pō'alima*, those patches whose product belonged to the chiefs), so that their favorite taros were available wherever they rested.[11] Cultivars preferred by the *maka'āinana* of a given place likely survived in family *lo'i* and isolated patches but retreated to the background of mainstream food exchanges as the tastes, preferences, and practices of more influential population centers (i.e., the communities of Lāhainā or Kona) held sway.[12]

A number of early foreigners influenced Hawai'i's agricultural practices with the introduction of ungulates (hoofed animals), alien plants, and planting practices. The delivery of domesticated European pigs was recorded in 1778 by Captain Cook. Additional introductions followed with later explorers, whalers, and missionaries.[13] Captain George Vancouver made a gift of goats to Kahekili in 1793, who sent them under *kapu* to Kaho'olawe (Social Science Research Institute 1998). In the following year, Vancouver successfully introduced cattle to the islands. These animals multiplied without restraint under a *kapu* system designed to protect and manage resources for abundance but that was ill prepared for the advantage this gave to invasive species. The *kapu* on cattle was lifted in 1830, but by then the damage to watersheds and *kalo* farms had already been done (Diong 1982). Within the district by district folios of the Bernice Pauahi Bishop Museum photograph collections are images from the later half of the nineteenth century of taro patches with fences to keep cattle out and upland forests, the source of their water, stunningly denuded in many places. Ranches took over dryland field system zones, including those of North Hawai'i and East Maui. Stone walls such as those found in Honua'ula district, Maui, were built to protect crops, sacred sites, and homes from cattle (T. Dunham personal communication, 2005; Maly and Maly 2005; Whelan and Menton 1988; Handy 1940). By the early 1900s, the islands of Kaho'olawe and Lāna'i had lost millions of tons of topsoil into the ocean from overgrazing.[14]

The period of foreign-driven sandalwood trade (~1800–1840), which coincided with the advance of hoofed animals into the mountains, forced *maka'āinana* to abandon their fields and their role as natural resource managers (Kamakau 1992). The initial flush of revenues and goods generated by this new trade encouraged Kamehameha to seek a register and seal from Britain to export taro in exchange for furs in 1810 (Hackler 1986), unaware that the sandalwood trade would lead to scarcity for many of his people. "Sandalwood harvesters were often gone for several days, sometimes for weeks, in the mountains collecting sandalwood. Many died of exposure and other misfortunes in the cold, often damp uplands" (Merlin and VanRavenswaay 1990, 52).[15] This extractive, out-of-*ahupua'a* trade resulted not just in neglected crops, food shortages, famine, and further watershed degradation, but also caused a break in the continuity of existing agricultural systems (Cottrell 2002; Merlin and Van-Ravenswaay 1990). The missionary William Ellis confirms conditions of poverty and abandoned fields during his journey around Hawai'i island in 1823 (Ellis 1825). He also noted occasional abundance (plantings of *kalo*, *'uala*, *mai'a*, and *kō* [sugarcane]) in some places, as well as new food crops, including melons, squash, onions, and pumpkins, vegetables that had been introduced by Cook and other early explorers. Such produce supplied foreign palates in the islands and markets abroad. The barter for goods with explorers was not so far removed from that of the exchange system in place among the peoples of the Pacific at the time (Collerson and Weisler 2007; Hansford 2007). But, it is here that the shift from crops as food, medicine, and *kinolau* (sacred body forms of the gods) to crops solely as commodity begins. The sandalwood trade became a significant catalyst in speeding the transition from food self-sufficiency and abundance to market driven agriculture.[16] By the 1830s, *kalo*, poi (pounded cooked taro), and *lū'au* (taro leaf) had a definable and growing market that supplied the centers of trade on each island, particularly the port of Honolulu.

It is in this same era that hundreds of new plant species were brought to Hawai'i to test their potential, many of which had devastating effects on agriculture and native ecosystems decades later when they reached critical mass.[17] The first large sugar plantations emerged in the 1830s, followed by consolidations of land and water in the hands of a few newcomers, and fifty years later by pineapple plantations.[18] The impact on traditional Hawaiian crops and planting practices became increasingly visible. The effect of the 1848 Mahele and the Kuleana Act of 1850, both originally intent on providing title to *maka'āinana* (commoners) as a means to increase native productivity on the land, were distorted by the time of the overthrow of the kingdom in 1893, resulting in broad dispossession of those lands (Preza 2010).[19] The effects of codified private property on *kalo* cultivation are more subtle than the overt loss of rights to land for the *maka'āinana*.[20] When a stream shifted course, a flood inundated a farm, or a set of terraces needed to fallow, the option to move plantings to another parcel of land or to access a stream at a different point became less of an option than it had been in the past.

Kō (sugarcane) was a common element in the Hawaiian garden.[21] In the Hawaiian context, it was planted as an amendment to staple crops in small plantings and as windbreaks with a built-in mulch source.[22] As with sandalwood, the multipurposed *kō* became a singular-purposed export crop for foreign businessmen.[23] On Maui, lowland forests and traditional Hawaiian agricultural field systems (wet and dry) from Hāna to Nahiku, the great district of Lāhainā, Kā'anapali, Kīhei, and most of Central Maui were supplanted by large fields of cane and pineapple on a scale that fit neither the landscape nor the available resources.[24] Massive soil runoff during heavy rains and dust-filled air when the winds blew became regular conditions, then as now. The majestic sand dunes that crossed the isthmus from Waihe'e to Kahului, Mā'alaea, and Mākena took back seat to tilled and irrigated sugar.[25]

An 1875 *Thrum's Almanac and Annual* records that sugar production increased from 300,000 pounds in 1846 to more than 24.5 million pounds in 1874 (Campbell and Mentin 1986) and exponentially to 260 million pounds by 1890 (Wilcox 1996). In 1865, the first immigrants were brought to Hawai'i to work labor in the fields and build the ditch systems that supported the plantations. Between 1850 and 1930, several hundred miles of irrigation ditches had significantly dewatered streams throughout East and West Maui to feed sugar and pineapple.[26] A similar story unfolded on each of the main Hawaiian

Islands. Within the span of ten years, from 1897 to 1907, an attempt at cane production on the island of Molokaʻi by the American Sugar Company Ltd. resulted in the overpumping of Molokaʻiʻs primary aquifer, which has never recovered to its original conditions (Coral Reef Assessment and Monitoring Program 2008; Cooke 1949). The company also introduced mangrove to the island in 1902 in an attempt to stem the flow of soil runoff from its fields. The tree colonized the coastline, altering the ecology of nearshore reefs and Molokaʻiʻs once prolific fishponds, as well as the nutrient dynamic between inland and coastal ponds.

Wetland and upland *kalo* systems were abandoned as ownership and control of land, and the waters that flowed through them, were removed from Hawaiian occupancy and authority. As the number of Hawaiian planters declined, Chinese and Japanese farmers who had arrived to work for the sugar companies took their place.[27] By the end of the nineteenth century, rice production had replaced *kalo* in the majority of *loʻi* lands on all islands. The breadth of *kalo* cultivation knowledge in practice a century earlier had greatly diminished, including many of the sophisticated mulching techniques of the past (Allen 2001; Olszewski 2000; Handy, Handy, and Pukui 1991; Handy 1940; Kalokuo-kamaile 1922; Iokepa, n.d.-a, -b; Queen Emma, n.d.; and Hawaiian language newspapers, such as *Ka Nupepa Kuokoa*).[28]

Within wetland taro field systems, traditionally wide banks between patches, often the site of other food and mulch crops, cordage, dye and medicinal plants, and even *imu* (underground ovens),[29] were narrowed to the width of a footpath in the habit of Asian rice fields. Where any number of taro varieties might be found in a single Hawaiian field or farm, Chinese and Japanese growers, while they did not abandon *kalo* diversity entirely, focused on the few cultivars most favored for market. The overall number of varieties dwindled to less than one-third of their original diversity of 300 to 400 cultivars as marginalized forms fell out of use (Handy 1940; Handy, Handy, and Pukui 1991; Whitney et al. 1939; Wilder, n.d.; MacCaughey and Emerson 1913–1914).[30] Some have suggested that by 1933 only thirty-two Hawaiian taro varieties survived as a market crop (Allen and Allen 1933; Lebo et al. 1999). In the lowlands, a pattern of combining small *loʻi* into larger units emerged, with the result that water through the system slowed and warmed. Taro root rot, restrained by cool, steady water flows, blossomed in the slower, warmer water temperatures of larger field systems.[31] Warmer water also influenced crop development rates. Based on present-day observations, it seems probable that *kalo* varieties known for their ability to remain in the patch for eighteen to twenty-four months in colder water or higher elevations required earlier harvest or may have struggled to grow under altered conditions.[32] Whereas before, the patches were turned and planted by many collaboratively,[33] farming became a business with the burden of labor falling to fewer and fewer individuals and families (Olszewski 2000; Salmoiraghi and Yoshinaga 1974). When cultivation practices and markets changed, as well as who was planting, the relationship between plant and planter also changed—from sacred and staple food to commodity crop. Caretaking the breath of Hāloanakalaukapalili, and the relationship between elder and younger brother, lay in the hands of a shrinking number of Hawaiian taro growers.

By 1900, feral ungulates and plantation agriculture practices had altered the landscape so radically that traditional Hawaiian farming practices were forever impacted. Wetland taro cultivation fell to roughly 1,280 acres, although 9,400 acres of rice kept the ancient systems open and somewhat intact (Smith 1902). Plantation ditch systems and acreage in sugar and pineapple continued to expand well into the 1920s. At the peak of the industry, an estimated 200,000 acres were planted in cane, yielding an estimated two billion pounds of sugar in 1932 (Wilcox 1996), but the applied practices came at great cost to the land. Growing methods for vegetable crops followed suit. A 1931 soils recovery bulletin noted that "farms on the slopes of Mt. Haleakalā, which supplied the California miners in 1849, and for years after, with wheat and potatoes, have for fifty years [since the 1880s] been unfit for agricultural crops and are now used as grazing land, because of soil losses. In almost every section of the Hawaiian Islands erosion is causing similar soil losses" (Zschokkee 1931, 9). Fishponds and wetlands that rimmed the shores of Moloka'i and embraced the shorelines of Central Maui were silted over by the runoff.[34] One of the least well-documented alterations to the sustainable elements of Hawaiian agriculture systems has been the disturbance of carefully engineered stonework and water distribution infrastructure, each one designed to the unique conditions of its location. Many have been dismantled or built on top of, including such famous sites as Kīkīaola (the Menehune Ditch) on Kaua'i (Taro Security and Purity Task Force 2009).[35]

The environmental damage from this new agriculture was so compelling that, as early as 1903, Hawai'i became one of a handful of nations with an established Department of Forestry in an attempt to reverse the impacts (Cox 1991).[36] Yet, the agriculture model in Hawai'i continues to submit that large-scale, export driven cropping systems are the most logical and economically viable despite evidence to the contrary, including the ecological collapse of farms in the US Midwest in the 1930s and the economic collapse of the family farms that replaced them, today.[37]

After World War II, the rice market in Hawai'i failed in the face of outside competition, and taro returned to the *lo'i*; however, by that time, so much water had been diverted for sugar that only remote streams remained untapped. A 1937 source indicated that wetland taro cultivation had increased slightly to 1,500 acres (Advertiser Publishing Co. 1937) but with rice no longer being cultivated, a significant percentage of *lo'i* lands were left to fallow. The shift in sociocultural ambitions and ideals that accompanied the end of World War II tipped the balance against a strong recovery of these former taro lands. In addition to the well-documented loss of native Hawaiian populations to disease, as well as changes in land ownership and water resources, native and nonnative taro farming families now urged their children to find new vocations rather than take up the backbreaking labor in the mud of the *lo'i*. As a result, the number of skilled taro farmers that were to become the next generation of growers declined even further.[38] Again, the introduction of new invasive species shifted the landscape and added to the already challenging work of maintaining a sustainable *kalo* production system. By 1940, the crayfish *Procambarus clarkii* had inflicted such serious damage on the narrowed banks of *lo'i kalo* that they leaked persistently or fell apart. Scientists recorded that crayfish "consume taro . . . [and] cause the draining of taro fields" (Bishop Museum 2011). Range grasses and nonnative tree species introduced as part of erosion control and improved grazing programs through the Hawai'i Forestry Department and the Soil Conservation Corps (which later became the Natural Resources Conservation Service) made their way downstream into *lo'i* and wetlands (Buchanan, Levin, and Pali 2010). California grass, napier grass, guinea grass, and barnyard grass, among many others, took hold and took over. In the 1970s, the aggressive cattail rush *Typha latifolia*, whose seeds have long-term persistance (Hawaii Weed Risk Assessment database 2011) found its way into the taro lands of Kaua'i and Waipi'o, Hawai'i.[39] The latest pest, the apple snail *Pomacea canaliculata*, arrived in the *lo'i* in the early 1980s and accounts for an 18–25 percent loss in taro production annually. Controlling this aggressive pest has increased labor for taro growers by 50 percent (Levin 2006). Most of these species were brought to Hawai'i with the intention of "improving" agriculture and/or developing an alternative food/economic crop.

Perhaps the most significant impact on agriculture practices in Hawai'i after World War II came with the transformation of chemicals of war into weapons of the field. Most commercial taro farmers abandoned traditional mulching practices by the 1960s in favor of chemical fertilizers. This substitution of historic mulching materials and methods, followed soon after by pesticides, has been a costly and extractive practice. Traditional mulches were soil health focused and managed soil texture, nutrients, and native mycorrhizae, rebuilt soil over the long term, and contained properties that helped keep taro pathogens in check. While less labor intensive, artificial inputs were plant yield focused, short-lived, expensive additions[40] that have contributed to a more than sixty-year decline in lo'i soil quality,[41] a legacy of chemical contamination in soils and wells (e.g., the Hamakuapoko wells in Maui), and nitrification of Hawaiian reefs (Hawai'i Department of Land and Natural Resources 2007).

The introduction of mechanized paddy tillers aided farmers in the management of larger fields with less manpower and replaced the hardy ponies and water buffalo that farmers had adapted a century earlier to work in the lo'i. The tractor had difficulty operating in the smallest fields but was an advantage in large patches for those skilled in maneuvering the machines in the mud. For most commercial taro growers, tilling remains the only mechanized portion of the growing cycle.

Between 1946 and 1980, land in commercial taro production (wetland and dryland) declined from about 1,020 acres to its lowest point of 320 acres (National Agriculture Statistics Service 1946–2013), a loss connected to the history described above but also to urbanization,[42] the claiming of taro lands by federal and state agencies as wetland wildlife habitat, and successive storms and floods that destroyed crops.[43] Hawai'i's changing population influenced the demand for taro and hence the acreage in production during the 1980s.[44] Poi consumption remained primarily within the Hawaiian communities, a demographic that made up just 9.3 percent of residents in 1970 (Begely, Spielmann, and Vieth 1981). With the introduction of apple snails to lo'i beginning in the mid-1980s[45] and an outbreak of taro root aphids in dryland plantings on Hawai'i island in the 1990s (Taro Security and Purity Task Force 2009), commercial taro acreage today hovers around 400 acres and 105 farms (National Agriculture Statistics Services).[46]

Generational attitudes about social reciprocity since the 1980s have resulted in a breakdown of traditional collective management of the 'auwai (irrigation canals) that fed taro farming communities. These main canals are the lifeblood of an entire system. Not only are there fewer farmers cleaning sections of 'auwai, but also fewer people willing to help others beyond the boundaries of their own farm.[47]

The outcome of reduced acreage in irrigated and rain-fed cultivation has been to discard additional elements of sustainable practice and disease management—long-term fallow rotations and wide planting spacings—in order to meet the demands of poi millers and other buyers. Traditional plant stock culling practices that eliminated *makua* (parent stock) and weak *huli* (taro tops) as a mechanism to reduce disease gave way to the needs of commercial growers for more planting material. Larger patches and closer planting distances required an estimated 11,000–27,800 *huli* per acre (Evans 2008, 42). Close plantings increased the chances of leaf blight spread. The presence of apple snails on many farms reduced, by necessity, the sharing of *huli* between growers as a safety measure to prevent further spread of the pest. During *huli* shortages, farmers had little choice over the continuous reintroduction of diseased stock that is still a challenge today. Most taro plants now carry some level of dasheen mosaic virus. Insect pests, including aphids, mites, and thrips, have grown increasingly pervasive on taro farms, slowing growth and scarring plants; the result of prolonged island-wide drought cycles that favor the pests. The search for better yields from ever weakening soils and plant stock has led to the introduction of taros from other regions of the Pacific and Asia and the development of disease-resistant hybrid taro cultivars, rather than a return to soil husbandry. In the short term, these introduced varieties and hybrids have assisted commercial growers to survive under the compounded, difficult conditions that currently exist on-farm, particularly for *lūʻau* leaf growers. Already, however, they are beginning to exhibit some of the same challenges as older cultivars, while soil quality, water, nutrient conditions, and *huli* stock vigor remain unaddressed.

If there is any evidence that the poi mills asserted a preference for what was planted, it is in the fact that by 1990 a single taro variety, the hybrid Maui Lehua, dominated commercial wetland fields, effectively turning *kalo* production into a monocrop with all its inherent risks. What little diversity remained in the fields, as evidenced by a 1964 taro variety density study (Bowers, Plucknett, and Younge 1964), had vanished for the mainstream market. Although a few small commercial and subsistence growers retained a broader range of varieties, consumers today have rarely tasted any other taro variety and consider Maui Lehua the measure for all other poi taros. The colors, tastes, and fragrances of the traditional Hawaiian taros are an elusive, foreign land to the majority of consumers, millers, and growers.[48]

As sugar has fallen out of production (barely 17,000 acres were harvested in 2010),[49] taro farmers have called upon the state and private landowners to support access to former taro lands so that current taro lands can rest. They have petitioned for the return of waters diverted by plantations to their streams of origin so that *kalo* can once again flourish as a staple crop for Hawai'i's consumers (Lukens 2010). Starved of water for 150 years, the potential for recovery of stream habitat, nearshore reefs, and this important food source remains in limbo even after thirty-five years of court proceedings. A new player in Hawai'i's agriculture narrative, the large-scale biotech seed industry, is now competing to retain the same hold on freshwater resources as the plantations— out of watershed and across whole islands against all common sense for ecological health or the state's food self-sufficiency.

The legacy of modern agriculture practices has been, until now, one of soil impoverishment, erosion, introduction of invasive species, loss of our native biota, indigenous crop cultivar diversity, and "seed stock" (planting material) quality, and the ongoing decline of the functional capacity of our watersheds and reefs. From this perspective, one *should* argue that an agriculture whose sole purpose is economic gain sacrifices natural resource sustainability.

But what was the course of agriculture productivity? From a Hawaiian nation of 300,000 to almost one million people (Kirch et al. 2007; Stannard 1989) that was food self-sufficient prior to the arrival of foreigners on these shores, Hawai'i has become a state of 1.3 million dependent on imports for an estimated 85 percent of our food (Page, Bony, and Schewel 2007)—a cargo cult, albeit a high-end one, by any other name. This includes imports of more than 1.8 million pounds of taro annually (Taro Security and Purity Task Force 2009, 42) in a land that was once famous for its sustainable production of this staple crop.[50] While there is no exact picture of the acreage once under wetland *kalo* (*lo'i*) cultivation, estimates of between 14,000 and 20,000 acres have been suggested historically (Evans 2008; MacCaughey and Emerson 1913–1914). Writing about water rights in Hawai'i, Hutchins (1946, 31) documents that, in the allocation of lands by 1850, "awards to the common people [were] estimated to have aggregated a little less than 30,000 acres" and that "Professor Kuykendall [of the University of Hawai'i] point[ed] out that nearly all of the awards . . . were very valuable for their own agricultural pursuits so long as enjoyment of the appurtenant water rights was assured." These parcels "consisted primarily of irrigated taro lands in the valleys, which were then regarded as by far the most valuable lands."[51] Gon (2010) and Ladefoged et al. (2009), using GIS modeling, have predicted at least 190 square kilometers (46,950 acres) of *lo'i kalo* and almost 738 square kilometers (182,364 acres) of *māla* or rainfed plantings (all crops) were potentially in cultivation with a small amount of lands that were both *lo'i* and *māla* (7 km²)—a total of 936 square kilometers or 231,291 acres of food-producing lands in the islands.[52] This surpassed land in sugar at its peak by more than 30,000 acres.[53]

On Maui, the districts of Na Wai Ehā (Waihe'e, Waiehu, Wailuku, and Waikapū), Ke'anae, Nāhiku, Lāhainā, Kā'anapali, and the upland rainfed south-facing slopes of Haleakalā, were once prolific food-producing lands, particularly in the forest zones of Honua'ula to Kaupō (Sterling 1998). As late as the 1950s, small family-run poi mills were still part of most taro-growing communities. The ledgers of at least one East Maui mill, the Ko'olau Poi Factory, which delivered poi all the way to Pā'ia and out to Hāna and Kaupō, recorded purchasing as much as 3,600 pounds of raw taro per week in 1944. Even accounting for fluctuations in production, upward of 170,000 pounds of taro served this single mill in that year, which in turn provided jobs, income, and food for hundreds of local families (Kanoa 'ohana, personal communication, 2011).[54]

The traditional system included thousands of farmers. Ladefoged et al. (2009) suggest more than 240,000 people would have labored in the production of food at the peak of Hawaiian cultivation, 29,700 of them in wetland taro systems. Planting and fishing were the primary activity of most families. Today, this trend has reversed: of the estimated 1.3 million residents in Hawai'i today, 7,500 were reportedly employed directly in farming and ranching in 2011 (Hawai'i Department of Agriculture)—less than 0.6 percent, and far below the national average of 2–3 percent of the country's 317 million population in 2013.[55] An even smaller percentage is growing food directly for our tables. Relying solely on Hawai'i Department of Agriculture statistics, only 370 acres of *kalo* (approximately 190 acres in the Hanalei National Wildlife Refuge) were in wetland production in 2013. When small, unrecorded commercial growers, subsistence farmers, educational and cultural *lo'i* are considered, perhaps 600 acres of *lo'i kalo* continue to be grown, still less than 1 percent of agricultural lands in the state (Taro Security and Purity Task Force 2009; National Agriculture Statistics Service 1946–2013).[56]

Spriggs (qtd. in Ladefoged et al., 2009, 7) suggests that wetland fields in Oceania were capable of producing between twenty-five and fifty metric tons of taro per hectare (10.1–20.2 mt/acre or 11.2–22.3 US tons/acre).[57] Relatively high yields are supported by earlier writers. MacCaughey and Emerson wrote that "a few square rods, under proper cultural methods, will continuously produce enough *kalo* to support a family. It is due to this great productivity that ancient Hawaii, despite its very limited area, was able to support a relatively dense population" (1912, 189). At the turn of the twentieth century, commercial farms "generally produce[d] from twelve to fifteen tons" per acre (189). The same source noted, however, that Chinese planters were "not in the habit of giving the land a rest" and "consequently, while the natives raised good *kalo* with many oha [young taro corms; offshoots], the Chinese get small *kalo* with almost no oha" (1913–1914, 22). This seems to suggest that while aggregate productivity was low, yields from the higher end of the scale applied to traditional *loʻi* production under the care of Hawaiian planters. While there has been a slight increase in taro acreage and yields in the last few years, in the big picture wetland taro production in Hawaiʻi fell from a projected 475,000 to 950,000 metric tons of poi taro (raw), or approximately 1.05 to 2.1 billion pounds annually in precontact times (Spriggs, qtd. in Ladefoged et al, 2009, 8), to just over 3.8 million pounds (1.7 metric tons) in 2013 (National Agriculture Statistics Service 1946–2013). Today, the yield in dryland *kalo* cultivation is estimated at between 18 to 20 US tons (18 metric tons; 40,000 pounds) per acre (G. Teves, personal communication, 2011) and falls within Spriggs' upper limits. Yet, looking at the mean for wetland production over sixty-seven years of data in Hawaiʻi shows a rate of only 8.2 US tons per acre (7.4 metric tons), peaking in 1966 at 11.2 US tons (10.2 metric tons) during the early years of chemical fertilizer use. Crop losses related to floods, apple snails, and birds notwithstanding, wetland yields since that time have been in relative decline. In the last five years, barely 4.5 US tons (4 metric tons) per acre were harvested with a twenty-year average of under 6.4 US tons (6 metric tons) (Hollyer et al. 1990; National Agriculture Statistics Services 1946–2013).[58] Some older taro-farming families describe significant losses in wetland productivity within the last thirty years, including such examples as harvests reduced from 100 eighty-pound bags to just thirty bags in a one-sixth acre field. Corm weight and quality has also diminished over time.

Photographic, agricultural, and ethnographic records depicting the quality of taro harvests from just two and three generations past in both wet and rainfed cultivation are elusive and complicated by images of poi pounding boards piled with *ohā*, the smaller corms preferred by many Hawaiians for their starchiness and flavor. Yet descriptions of corms weighing ten pounds or more and an average eight to ten corms to a one-hundred-pound bag from rainfed and irrigated cultivation have been passed down by *kupuna* (elders) on each island.[59] One farmer from Puna recalls disbelief when his grandmother described corms reaching fifteen pounds or more until he was able to duplicate it in 2005 (J. Konanui, personal communication, 2005). A farmer on Molokaʻi remembers a story from an old-time grower who produced a dryland *kalo* that weighed sixty-five pounds using a pit method of planting (G. Teves, personal communication, 2011). Queen Emma, in a manuscript from the mid-1800s, provides the quintessential description, writing, "I have produced kalo which averaged twenty-two inches in length and the same in circumference when it was cultivated under my own eye" using the time-tested mulching practices of old Hawai'i. At peak density, this was *kalo* of perhaps twenty pounds or more. Today, the average weight of a taro corm in wetland production is two to four pounds,[60] and the percentage of *loliloli* (overripe), *palahe'e* condition (rotten),[61] and crayfish or apple snail damage can exceed 50 percent. This is a stunning change in food self-sufficiency.

DEFINING SUSTAINABLE AGRICULTURE IN HAWAI'I: PARALLEL AND DISCORDANT DIALOGUES?

At least two discernible courses in the sustainable dialogue can be found— political-economic and on-farm. They hold in common a goal of economic benefit (profit). How "economic benefit" is defined shapes the meaning of sustainable agriculture on the ground and in the legislature in the state of Hawai'i.

The underpinnings of sustainability definitions begin with the nation-to-nation dialogues of the Brundtland Commission in the 1980s. The most well-known and perhaps most generic interpretation was adopted by the commission in 1987: "Sustainable development is development that meets the needs of the present without compromising the ability of future generations to meet their own needs" (World Commission on Environment and Development 1987). Just a few years later in 1991, the International Union for Conservation of Nature, United Nations Environment Program, and World Wildlife Fund suggested that natural resource extraction had its limits: "Sustainability is improving the quality of human life while living within the carrying capacity of supporting eco-systems" (IUCN, UNEP, and WWF 1991). Powerful nations and industries shaped the first definition; the second was a response by environmental groups who felt the first had left out a key principle of finite resources. Laenui (1993) called both these definitions of sustainability a "domination approach"; one predicated on the assumption that the purpose of the environment is to "support, sustain and be a resource for humankind." It asks, "how much environmental degradation should we tolerate," and, "how do we maintain that level of tolerance without overstepping that boundary?" In contrast, from an indigenous approach of reverence for the earth, one which lacks ownership but is full of belonging, and which often "treat[s] the interrelationship with the environment as nothing less than spiritual work of the highest degree," Laenui and the Hawaiians interviewed suggested the questions change from, for example, "how do we protect our ocean as a valuable resource?" to "how do we protect this womb, this sacred place of creation . . ."

Three decades later, no single definition since the Brundtland Commission has been fully supported. Internationally, small-scale farmers and indigenous peoples throughout the world have continually advocated for more balanced, less environmentally demanding agriculture practices with varying degrees of success (through UNESCO, FAO, and other forums).

Until recently, discussions of sustainability and sustainable agriculture in Hawai'i have paralleled the domination dialogue. Both globally and locally, global warming, depletion of natural resources, and challenging economic conditions have been a positive catalyst for change. To fully understand the somewhat disconnected path of "sustainability" and "sustainable agriculture" in Hawai'i's political mainstream, it is important to consider the concept of "diversified agriculture" that was reintroduced in the 1980s and often used to imply or measure sustainability in the state, in contrast to its meaning in 1902. A critical question should be asked: Is "diversified" a substitute for "sustainable"? And if so, at what level?

Stanford Dole, first territorial governor and ironically one of the architects of the overthrow, made clear the relationship of diversified agriculture to small farms "in distinction from sugar-planting interests" when he wrote, "I need hardly remind you of the extent to which our political future depends upon the growth of a farming class in these islands, living on and making their living from their farms. If we fail in this, and [sic] the agricultural work in the Territory shall be confined to large estates by a floating element of cheap laborers having no interest in the soil . . ." (letter to the Farmers' Institute in USDA Office of Agriculture Experiment Stations, Smith 1902).

Former governor and lawmaker George Ariyoshi recounted that when the tariff acts that protected sugar and pineapple pricing in Hawai'i lapsed, a new direction in agriculture had to be found. He wrote, "The situation was particularly painful because [after 150 years] our sugar industry was highly efficient. . . . Looking for alternatives to large-scale agriculture, the state government undertook a multifaceted program of nurturing the diversification of agriculture" (Ariyoshi 2003). Few at the time questioned whether the practices inherent in the industry were sustainable or not. Small farms, particularly for crops like taro, were inconsequential to the equation. At the highest levels of state government, agriculture institutions, and industry, the pursuit of diversified agriculture was still primarily about maximizing yields and sustaining or increasing the revenues derived from agriculture (Ariyoshi 2003).

After twenty-five years, diversified agriculture is still being measured by the diversity of crops produced, sold, and exported from Hawai'i, as a whole[62] but has notably remained focused on crop substitution (e.g., seed crops, papaya, or coffee for sugar and pineapple) in the field. In 2009 the National Agriculture Statistics Service listed only 6,175 acres in diversified agriculture (a selection of fruit, vegetable, and melon crops, including exported crops), less than 0.55 percent of all agricultural lands in the state (2007 statistics). Practices such as agroforestry, intercropping and organics, crop biodiversity (a multitude of varieties of a single crop), or a multitude of crops within a single field or annual cycle are found primarily on about 5,350 small family farms of less than ten acres each.[63] For the largest agriculture producers, the size of the fields did not shrink, nor did the focus on exports and eliminating competition.

In 2006, the understanding of what sustainable agriculture could be in Hawai'i appeared to have shifted—at least in writing. Equating sustainable agriculture with "ecology-based farming," the University of Hawai'i College of Tropical Agriculture wrote that sustainable farms should be "capable of maintaining their productivity and usefulness to society indefinitely" and that it should be "resource-conserving, socially supporting, commercially competitive, and environmentally sound" (Smith and El-Swaify 2006, 1–2). The college acknowledged the "unintended consequences" of modern agriculture and the difficulties of being a farmer and detailed a series of both conventional and organic supporting practices. Three primary goals were considered: (1) providing a profitable farm income based on a productive enterprise, (2) promoting environmental stewardship, and (3) promoting stable, thriving farm families and communities. In 2013, there is encouraging exploration at the university into the use of organic practices such as indigenous microorganisms[64] and more support for organic agriculture, including recognition that it has a critical role to play in the big picture of sustainable agriculture. The challenge is in the transition from publication to "promotion" to implementation across a broad array of crops and field/industry sizes—and also to policy makers. Agriculture development in the state still predicates decision making on large-scale production and export-driven or high revenue-generating markets. To illustrate, a recent private analysis of the potential of small-scale, diversified agriculture in Maui wrote that "while the Maui Island market is significant, it is comparatively small . . . [with a] *de facto* population of 156,170 residents and visitors." The report concluded that because of the costs of shipping to Honolulu and the mainland, Maui farmers could not compete and that "[local] consumption volumes are too small to support large, efficient farms" (Decision Analysts Hawai'i, Inc. 2009, 7, 17), despite a locally grown food deficit of 85 percent.

In the last decade several statewide, county-level, and island-based planning processes addressed the issue of sustainability and sustainable agriculture. Three of the most recent plans provide insight into how Hawai'i's views of sustainable agriculture are evolving. In 2007, after a yearlong process of community input, the Hawai'i 2050 Sustainability Task Force summarized the public's vision of what "sustainability" meant: "To some, it was about protecting the environment. To others, it meant creating agricultural self-sufficiency and self-reliance—living in a self-contained system. Others viewed it as a matter of economic resilience" (Hawai'i 2050 Sustainability Task Force 2007). A plan involving so many people and agendas has its challenges; however, it made note of one important thing: "In a sustainable society, *systems replenish themselves* [italics added]. They don't rely on the consumption of economic, social and environmental assets for progress." And yet, the 2050 plan's definition of sustainability fell back to the more common discourse of human-centered, "balanced" consumption and the definition of the Brundtland Commission:[65] divvying up the pie, but not replenishment.

The *Molokai: Future of a Hawaiian Island* plan (2008, 13) recognized that it is important to protect Hawai'i from "outside interests that conflict with the island's values." The plan states: "It is imperative that the agriculture and aquaculture sector on Moloka'i thrive apart from the notion that we should import all that we consume. With only 2 weeks of food inventory in the state, attaining food self-sufficiency is a major priority—not only for Moloka'i, but for all of Hawai'i" (13). In discussing the future of agriculture, this small community recognized the abundance of their past. "In ancient times, the people of Moloka'i were renowned for their ability to produce abundant quantities of food. In honor of the great productivity of the island and its surrounding ocean, Moloka'i was frequently referred to as 'Āina Momona (abundant land). Through careful stewardship, Moloka'i's people were able to maintain a sustainable and self-sufficient food supply for thousands of years" (13). The Moloka'i plan recognized and defined the importance of subsistence as an essential part of agricultural sustainability discussions for its residents: "Subsistence is an important part of Moloka'i's hidden economy and a key to food sustainability and self-sufficiency. This economy thrives on Moloka'i today, as 38% of our food is acquired through subsistence activities. The skills needed to perpetuate a subsistence economy are based in cultural knowledge, traditions, and rights. Therefore, subsistence needs to be recognized, protected, and enhanced" (5).

A year later, the Kohala Center issued the *Hawai'i Sustainability Primer: Step by Natural Step*, which summarized that success in a sustainable society meant that "we are not systematically undermining nature's ability to provide the natural resources and ecosystem services upon which all life depends" (Natural Step 2009, 7). The *Island of Hawai'i Whole Systems Project Phase I Report* (Page, Bony, and Schewel 2007), a companion to the *Primer*, indicated that a wide variety of locally grown foods for the island captured a significant percent of the market, but it becomes clear that "local market share" is not the same as local consumption of locally grown produce (just 7 percent).

These plans affirmed that "diversified" and "subsistence" were two terms that should be examined more closely in illuminating the meaning of sustainable agriculture in Hawai'i. This was reiterated in July 2010 by participants in the Maui Agriculture Design Conference,[66] who suggested that these core words, including "agriculture" (commercial and subsistence) needed to be redefined as the context of food production and land stewardship in Maui changed and that definitions should be "local and meaningful." Like the Moloka'i plan, they concluded that understanding the history and context of place, especially within each *ahupua'a*, and restoring cultural connections, specifically traditional crop systems and practices such as *kalo, 'uala,* and fishponds, were key elements in revitalizing whole systems agriculture, including caring for the *mauka*-to-*makai* (mountains to the sea) relationships of water resources. The core and strength of the network that was envisioned was based on a foundation of restoring *'āina* (land, inclusive of all its elements). This echoes Hawaiian concepts of relationship with the environment.

More recently, the Taro Security and Purity Task Force legislative report *The Taro Lives; Abundance Returns to the Land* (2009) addressed the issue of sustainable agriculture wholly within the context of *kalo*. The task force met with taro farming communities on all the main islands over the course of a year to record the complexity of issues facing taro farmers and make recommendations for its survival. In defining "the taro farming lifestyle" the task force wrote, "Taro farming communities are a critically important repository of traditional knowledge and practice and a model of sustainability. . . . the taro farming lifestyle is holistic; when a farmer tends the taro he or she is connected to everything else—the lands, the streams and the reefs" (16). The report recognized the need for ecological, family, crop, and economic well-being, and it called for consumers, landowners, agencies, council members, and legislators to take up their *kuleana* (responsibilities) to ensure that *maka'āinana* (in this case, taro farmers) and the lands on which the *kalo* grows "are cared for in turn" (17). For a taro farmer, commercial or subsistence, whether articulated or not, reciprocity and *waiwai* (wealth) are intimately linked. This often missed component of the sustainability equation, that those who do not grow food must create and protect the space in which food is grown for them, brings us back full circle to the original agreements between *kalo* and man.

These plans illustrate that definitions of sustainable agriculture at play in Hawai'i are only beginning to shift us from use (our share of the pie, even if that is sharing with the environment) towards abundance (making more pie). All of them use words we are familiar with in Hawai'i as guiding principles—*aloha 'āina* (love of the land), *mālama 'āina* (care for the land), and *ahupua'a* management (mountains-to-the-sea integrated natural resource management)—that sound well enough but are not necessarily applied in day-to-day decision making because we have been so long away from it. Or, we are tripped up by too many conflicting agendas. What can be observed is that as discussions of sustainable agriculture become more localized (smaller communities with similar understandings of place), they begin to define what practice looks like in more detail and with much greater agreement.

The sustainability dialogue has not moved past "resource use" to "restoration," yet. The pieces are there, but a caution is needed. A "sustainable agriculture industry" (revenue generation) is not the same as "sustainable agriculture" (resource rejuvenation). A recent Food Summit convened by the Hawaii Farmers Union United with lawmakers, agencies, and farmers in January 2014 illustrated the tension that continues between these two perspectives and a search for *pono* (balanced) ground. To shift from general policy statements to on-the-ground, viable implementation requires thoughtful consideration of the *hi'ohi'ona* (elements) and the *mole* (foundation) that *together* arrive at that place where sust'āina-bility sits.

NOTES

1. Abbott (1992, 5) counts twenty-three of the twenty-nine plants introduced to the Pacific from Oceania in prehistoric times as established canoe plants in Hawai'i. White (1994) describes twenty-four species. An additional species of bamboo (*Schizostachyum glaucifolium*) was considered part of the Polynesian plant assemblage by Wagner, Herbst, and Sohmer (1999). Sohmer and Gustafson (1987, 15–17) suggested there was general

agreement on twenty-seven plant species, with six others "not universally agreed to by ethnobotanists." Basing the list on the work of Abbott and others, twenty-five plant taxa seem to be the current consensus.

2. 'Ulu (breadfruit) was also a canoe plant, but its arrival in Hawai'i is placed at roughly 1200 AD, somewhat later than other canoe plant crops (Zerega, Ragone, and Motely 2004; Morton 1987). As a primary staple crop, 'ulu plays a smaller role in Hawai'i than in other islands of the Pacific. Kepler and Rust (2011) argue that mai'a held a more significant position as an ancient food crop in Hawai'i than has been previously considered.

3. Pa'i 'ai is almost waterless, pounded cooked kalo (taro), the stage before kalo becomes poi.

4. Descriptions of these systems can be found in numerous sources, including Kagawa and Vitousek (2012), Kirch (1994; 1985), Handy, Handy, and Pukui (1991) Kelly (1983), and Handy (1940); with additional references in Kirch (1994). Today, their remnants—stone walls, 'auwai (irrigation canals), and even kalo—can be found in many valleys on all the main islands except Kaho'olawe. As the Taro Security and Purity Task Force 2010 report to the legislature (2009) describes, much of Honolulu and urban centers on each island were built on top of these once prolific field systems.

5. Hawaiian land divisions were both natural resource based and political with boundaries that frequently followed topographic features such as ridgelines, but also crossed multiple ecological zones.

6. "Precontact" is a commonly accepted reference to the arrival of Captain Cook in the Hawaiian Islands in 1778.

7. Albeit, often recorded through the lens of writers whose understanding of what they were seeing was limited and influenced by foreign worldviews.

8. This is also observed in language with regional dialects. Examples of place-based varietal names for kalo such as Pololu and Papapueo are reflective of this dynamic.

9. Hawai'i and Maui were relentlessly at war between 1759 and 1795 (Reeve, n.d.; Kirch 1994). The names of kalo varieties, such as 'Owene and Manini 'owali, hint at the last remains of a harvested field.

10. Kalo varieties such as Nawao, described as having a domestic and "wild," or inedible form (Andrews 2003, 418; Pukui and Elbert 1986, 263), may have been the result of the movement of food into the uplands or from the uplands into the lo'i, under such conditions.

11. The "ali'i taros," those that make a red poi as they ferment, may have become commonly known in this way. Handy (1940, 12) writes, "There was undoubtedly more [localization] in ancient times, though I feel sure that the popular taros such as Lehua and Pii Alii, of which the chiefs were fond, and medical and ceremonial taros . . . were cultivated throughout Hawaii."

12. Fashion and fad can be seen in ornamentation and tattoo (Allen, 2006); it is reasonable to consider that foods may have been similarly influenced. The displacement of Silversword and Ha'akea kalo varieties by Maui Lehua in Ke'anae-Wailuanui taro districts is an example from more modern times.

13. The small-statured Polynesian pig (likely Sus scrofa ssp. vittatus) arrived with the first Hawaiians. Foreign ships released the much larger domestic pig (Sus scrofa) among the Pacific Islands, perhaps as early as 1513, to establish a ready food source for future ex-

peditions and "improve" the size of native pigs, with tragic results for the endemic forests of the Pacific (Diong 1982).

14. The island of Kahoʻolawe continues to lose approximately 1.9 million tons of soil annually primarily as a result of this earlier damage, despite the bombing and live fire training that occurred on the island between World War I and 1990 (Social Science Research Institute 1998). Molokaʻi's west end followed suit almost a century later with uncontrolled populations of deer. The rapid expansion of deer throughout Maui as well as recent introductions to Hawaiʻi Island, will put to the test our ability to restore our watersheds and the soils that hold our agricultural future.

15. Egged on by merchants who extended easy credit and luxury items at grossly inflated prices to Kamehameha I and *aliʻi* that followed, the sandalwood trade and the Sandalwood Tax of 1826 pushed *makaʻāinana* into poverty. The tax levied to pay off the nation's acquired debt (reported to be $500,000 in 1826) required every man to provide one half of a picul (66.66 lbs) of sandalwood to the governor of the district to which he belonged in a short ten months to generate revenues for payment of foreign goods (Merlin and VanRavenswaay 1990).

16. The word "subsistence" is often used to describe indigenous agriculture (for a discussion on the political, economic, and legal implications of this label in the present day, see Taro Security and Purity Task Force 2009); however, it is clear from the many descriptions of "gardens" and "plantations" in historic records (e.g., Handy 1940; Ellis 1825) that, from the Hawaiian perspective (and through the eyes of early visitors), a great deal of thoughtful abundance was associated with the agricultural systems of Hawaiʻi. "Subsistence" did not fit the circumstances.

17. The concept of noxious weeds was known to Hawaiʻi prior to the turn of the twentieth century, as referenced in the Hawaiian Forester and Agriculturist of 1904 (Board of Commissioners of Agriculture and Forestry 1904); however, the current understanding of invasive species did not become a significant factor in Hawaiʻi conservation decisions until recently. Even in 2014, the conflict between the perceived economic opportunities of new agricultural crops currently alien to Hawaiʻi and the need for a much broader, long-term critique of their impacts on existing agriculture traditions and to island ecosystems remains.

18. A distinction is made between the sugarcanes grown by Hawaiians and the plantations. The first sugar was milled and processed on a small farm in 1802 (Lānaʻi), but the more familiar, large plantation model emerged later with the establishment of the Ladd & Co. sugar plantation on Kauaʻi in 1835. The first pineapple plantation was established in 1886.

19. The Kuleana Act intentionally limited land awards to areas in "actual cultivation by each claimant" rather than favoring *aliʻi* claims for larger portions of land (Preza 2010, 105).

20. The western concept of privately owned lands did not exist in Hawaiʻi prior to 1900. Following swiftly on the heels of the overthrow (1893), the Organics Act (1900) shifted land tenure to foreign hands and land and water resources were consolidated into large holdings for production. For an in-depth understanding of the implications of land redistribution under the Mahele, see Andrade (2008) and Preza (2010).

21. An estimated forty to sixty Hawaiian cultivars of sugarcane have been identified that were developed prior to the introduction of foreign sugar stock and hybrids created

by plant breeders at the Hawaiian Sugar Planters' Association (N. Lincoln, personal communication, 2010; Schenk et al. 2004; Moir 1933).

22. Such as the famous *kō kea* windbreaks in the dryland planting systems of Kohala, Hawai'i.

23. *Kō* was used by Hawaiians in medicine, ceremony, *hana aloha* (love magic), weaving, games, and as a complement to *'awa* (kava). It was an ingredient in food dishes, a snack or energy food, and occasionally a dye plant. Sandalwood was used for perfuming and dyeing of *kapa* (tapa); in the making of musical instruments, household, and ritual items; in medicine; and as firewood.

24. These areas were either covered with wet forest (Hāna to Nahiku) or too dry to support cane without the extensive diversion of water to fields from outside the district, as in the case of Central Maui.

25. The lithified sand dunes of Maui are now one of its rarest habitats. Remnants of these 200-foot-high dunes can still be observed at the Waihe'e Coastal Dunes and Wetlands Refuge and in central Kahului. They continue to protect the *'iwi* (bones) of warriors who battled under such *ali'i* (chiefs) as Kahekili, Kalaniōpu'u, and Kamehameha for the right to reign over Maui. The remainder, along with the lowland cinder cone Pu'u Hele, were graded or mined for construction and beach replenishment, forever changing the landscape.

26. This included redirecting water from ancient bogs on the heights of West Maui into irrigation systems that would feed Honolua Ranch and the new agriculture industry in Lāhainā (R. Hobdy, personal communication, 2013; D. T. Fleming Arboretum at Pu'u Mahoe).

27. Directly tied to a rapidly shrinking native population due, in large part, to foreign disease and the loss of family lands.

28. The most familiar English-language cultural references, including David Malo, John Papa 'Ī'ī, Abraham Fornander, and Samuel Kamakau describe just fragments, as 'Ī'ī's title *Fragments of Hawaiian History* (1959) suggests, of the complexity of the Hawaiian world. Mookini (1974) illustrates the literacy of the Hawaiian nation, listing 106 newspapers, newsletters, and journals published between 1834 and 1948. *Ka Nupepa Kuokoa* (The Independent), one of the longest running from 1861 to 1927, became an important forum for Hawaiians to share their rapidly disappearing cultural knowledge. With the resurgence of the Hawaiian language, details from native language newspapers are only now beginning to hint at the true depth of information that was lost (Nogelmeier 2010). These sources illustrate that by the 1860s, readers were being called to remember the names of the *kalo,* the *'uala,* the *kō,* and the practices of cultivation before they disappeared and articles about *kalo* in the Hawaiian newspapers were already being written in past tense (Iokepa n.d.-a, -b).

29. As found at Limahuli, Kaua'i ('Onipa'a Nā Hui Kalo 2003, 121).

30. Estimates are based on evaluation of a number of sources and discussions. Whitney, Bowers, and Takahashi (1939) is considered the standard authority of cultivar identity from the 1930s; other sources indicate the compilation is incomplete. The *Bulletin* documents just eighty-four varieties, of which sixty-nine were Hawaiian cultivars, one was of unknown origin ('Iliuaua), and fourteen originated in the Pacific Islands or Asia (Japan and China). Wilder (n.d.) in an unfinished manuscript described only forty-eight varieties (sixty-five names) and suggested a potential of 200 varieties; Handy (1940) anticipated

250. MacCaughey and Emerson (1913–1914) provide vague descriptions for 266 names. The Bishop Museum (2007) Hawaiian *Kalo* Database includes a list of 664 names and synonyms, including introduced Pacific and Asian varieties.

31. The first English-language scientific report on taro root rot is Sedgwick (1902). Warm, insufficient water continues to play a significant role in taro disease levels.

32. The clonal nature of *kalo* may also point to epigenetic factors in the survival of some cultivars over others.

33. See descriptions in Malo (1898), ʻĪʻī (1959), and Kamakau (1898 [1992]).

34. By the 1970s, some fishponds had become so unrecognizable as to be considered solid land suitable for building.

35. There is little protection for ancient agricultural systems in Hawaiʻi. While state law now requires archaeological documentation before a site can be destroyed, nothing requires that they be protected in place. The details of how a system functioned, critical to understanding how water behaved in each place, remain unrecorded (Taro Security and Purity Task Force 2009). That can only be learned through protection and rehabilitation of the sites, but few landowner incentives exist to do so. An example of what these ancient designs have to teach us is in the story of *loʻi* rehabilitation at Wailau (ʻOnipaʻa Nā Hui Kalo 2003, 119).

36. The earliest formal forestry agencies appear to be the India Forestry Service created in 1864 and the Royal Forestry Department of Thailand in 1896, followed by Malaysia in 1901. A Division of Forestry was established in 1881 in the United States that evolved into the US Forest Service in 1905. England did not have a Forestry Commission until 1919. The creation of the Territory of Hawaiʻi's Forestry Department is superseded by the passage of an "Act for the Protection and Preservation of Woods and Forests" in 1876 by the sovereign Hawaiian government, fully fifteen years before the US Forest Reserves Act of 1891. The nation of Hawaiʻi was overthrown in 1893. In 1903, when the Forestry Department was established, the islands were a territory of the United States and did not become a state until 1959. Observations of forest health issued soon after the department's creation have much in common with present conditions (Giffard 1913; Hall 1904). It is interesting to note that prior to World War II, the sugar companies that owned the ditch systems were required under this new forestry department to outplant thousands of native and nonnative tree seedlings to reforest the watersheds.

37. While it can be argued that corporate agriculture appears to be experiencing an upturn in economics lately, conventional family farms in the mainland United States have failed at tremendous rates in the last twenty years under a heavy burden of debt from the cost of machinery, fuel, fertilizer, and seed.

38. With the advent of the "modern" era, a culture of shame for speaking the Hawaiian language and practicing old traditions that began subtly with the arrival of missionaries eighty years prior came to a head. By the 1950s, so many *mānaleo* (native speakers) had hidden their ability to speak Hawaiian and their knowledge of the past from their own children and grandchildren that the link was almost severed. With the exception of a few taro farming families, it took a cultural renaissance in the 1970s to revive the connection and for young Hawaiian and non-Hawaiian students to return to the *loʻi*. One bright spark occurred in 1939 with the printing of *Bulletin 84: Taro Varieties in Hawaii* by the University of Hawaiʻi. The book inspired a small number of taro growers, including Harry Masashi

"Cowboy" Otsuka with the help of his Hawaiian wife, to search out and plant the ancient Hawaiian taro varieties once again.

39. Aggressive chemical treatment reduced populations of this weed to manageable levels; however, restrictions on herbicide use in freshwater habitats have resulted in recolonization. Limited alternatives, and millions of windblown seeds, bodes ill for the future of *lo'i* and wetlands on all islands.

40. Nationally, fertilizer costs climbed 264 percent between 2002 and 2008 and another 28 percent in 2011 (National Agriculture Statistics Service 2011). Fertilizer expenditures represent the single largest cash output for taro farmers statewide due in part to extra-high shipping costs (no accurate figures are available; estimates of 25–30 percent have been suggested by individual growers). Pest control represents an additional 30 percent of total inputs (80 percent of that in labor) (Ooka and Brennan 2000).

41. The regular addition of organic matter to *lo'i* soils over hundreds of years was no longer part of farm management, and hence, little new soil was being formed. The carbon addition from organic matter helped keep wet clay soils easy to work. Written records (published and unpublished) hint that before the advent of so many wetland weed species in the taro patch, traditional mulches may have been composted prior to incorporation into the mud and planting. Today, the most common organic addition is taro leaves and stalks at harvest (often including diseased portions of the plant or weeds during the growing stage).

42. As early as 1911, Hawaiian newspapers noted concern for the loss of taro growing lands. In an article titled "Gone Will Be the Waving and Fluttering of the Taro Leaves," a writer expresses alarm that the taro growing lands of the Bishop Estate in Honolulu may be replaced with houses (*Kuokoa Home Rula* August 18, 1911; trans. H. Pellegrino). In the last decade, several parcels in Keʻanae-Wailuanui, Maui, were sold, and homes permitted to be built within the footprint of traditional *lo'i*, disconnecting them from surrounding taro patches and interrupting ditch systems.

43. From 1946 on, declines in taro production can be correlated with the occurrence of major storm events and after 1984 the arrival of apple snails, with relative consistency through 2009, after which evidence is less clear and may indicate a plateau in yields (Levin, unpublished notes). Production can take an average two years to recover after flooding, but data suggest a sometimes positive response to the influx of new soil and nutrients in less time.

44. A significant period of influence is noted with the influx of workers from China and Japan in the late 1800s, whose main staple was rice. The second period was influenced by the later wave of Filipino plantation workers, followed by a more modern generation of local residents whose food tastes were changing. Whereas the taro flour market had a following during the first half of the twentieth century, that disappeared, along with canned poi, after World War II.

45. The earliest documentation of apple snails in Hawaiʻi is from the late 1970s through the aquarium industry; however, there is no evidence that aquarium snails (primarily from urban homes) were the source of known introductions of the snail to rural wet taro fields on Maui around 1983–1984 and on Kauaʻi several years later (Levin 2006).

46. Chinese (Bunlong) table taro production, the dominant dryland taro, dropped from 250 acres to 10 acres between 1992 and 2004. Taro acreage is significantly higher

(potentially by 50 percent) when subsistence and the smallest commercial farms are included.

47. Today, only a small number of taro farmers within each valley still know the entire 'auwai system of their area. Many ditches have fallen into disrepair for just this reason. The main 'auwai often runs a distance too far for a single farmer to maintain. Collective management also helped to make sure each farm got its fair share of water. As the social and physical systems break down or have been abandoned, farmers have shifted to individual PVC pipe intakes on the streams.

48. In 2013 this is changing, thanks to the perseverance of those protecting the Hawaiian *kalo* cultivars, a growing number of taro farmers planting the ancient varieties, a revival of the tradition of making hand-pounded *pa'i 'ai* and poi, and Act 107, which legalized the sale of hand-pounded *pa'i 'ai* and poi to consumers for the first time since the adoption of modern food safety laws in Hawai'i.

49. Hawaiian Commercial and Sugar Company listed 35,000 acres in production in 2013 in Central Maui, the only place that sugar continues to be grown in the islands.

50. Statistics are estimates from 2006. The number has likely surpassed 2 million pounds, equivalent to more than half of local production. Limited access to bill-of-lading information since 9/11 has made it difficult for the Hawai'i Department of Agriculture to gain accurate data on imports. Speaking to local self-sufficiency, the Taro Security and Purity Task Force (2009, 42) states it "finds no logical reason why we should continue to import any type of taro to meet local needs."

51. This number does not include taro lands that were located within royal patents issued to the *ali'i*, crown lands, or government allocations, which totaled 1.5 million, 1 million, and 1.5 million acres, respectively. It is interesting to compare this to the roughly 30,000 acres of *lo'i kalo* that are estimated could feed Hawai'i's 1.3 million people the equivalent of 2.5 cups of *kalo* per day for an entire year, today (Reppun 2010).

52. While the model may have over- or underestimated growing areas based on the parameters selected, particularly for elevation and slope, and does not account for changing patterns of rainfall and water resources over time as forests were depleted and streams diverted, it provides us with one of the most well-reasoned assessments of the capacity for food self-sufficiency in Hawai'i that we have today.

53. In 2009, an estimated 1.12 million acres were in farmland in Hawai'i, including sugar, pineapple, ranching, and all other reported crops; sugar harvests accounted for only 22,200 acres or 1.98 percent. In 2010, this was reduced even further to just 17,200 acres of sugar harvested or only 1.5 percent of total agriculture acres (National Agriculture Statistics Service, 1946–2013).

54. Numerous small poi mills existed on each island. The years 1950–1960 were a pivotal time for the closure of many of them on Maui. Aloha Poi, still producing in Wailuku, is the last of this era. Families who grew *kalo* and managed to stay in farming were forced to deliver produce to more centrally located mills or ship to Honolulu. When the 1946 tsunami and later a major flood in 1964 swept through Keanae-Wailuanui, homes and fields were lost. For some, it represented the end of an era.

55. Reported agriculture labor in Hawai'i appears to have declined in 2013 to just over 6,000 workers, despite a slight increase in the number of sugar and pineapple jobs (800).

56. Estimates of subsistence acres are based on networks of taro growers statewide. Hawaiʻi Department of Agriculture Statistics Services data record a total of 400 acres harvested from commercial wetland taro farms.

57. Spriggs appears to derive his estimates at least in part from dryland taro production statistics in the Asia-Pacific region. FAO reports on taro also draw heavily on information from dryland cultivation. However, given historical descriptions of productivity in Hawaiʻi, these numbers retain some validity.

58. The average number of acres in commercial production has remained relatively stable at 400–450 acres per year since 1960, suggesting that there may be a declining response to conventional fertilizer applications.

59. Kamakau (1898 [1992], 37) verifies this, describing in the 1890s that "when a planting hole which showed no leaves was opened up, there would be a mound of taro heavy enough to weigh down a man or woman. . . . Most of the [sweet] potatoes and taros of today are not even half the size of those of the old days."

60. While recent hybrids can reach greater weights, this is not the norm for most growers.

61. Taro farmers often use this term in relation to corms with soft rots caused by fungal pathogens.

62. Accurate data on diversified agriculture is difficult to find or correlate; some statistics include the seed industry, ranching, and pineapple in evaluations. Page, Bony, and Schewel (2007, 19) define "diversified agriculture" as "crops other than sugarcane and pineapple."

63. Available National Agriculture Statistics Service and local statistics are relatively outdated; data here is from 2007. Up-to-date data is challenged by diminished agency funding for agriculture statistical services and data access-limiting aggregation practices that hinder accurate or in-depth analysis. Both Hawaiʻi and national statistics also rely on self-reported information. Small farms and small farm employment statistics are significantly under reported.

64. "Indigenous microorganisms" is a term and practice used and taught by Master Cho, founder of Korean Natural Farming, but the concepts and mechanisms are not unknown to older agricultural practices worldwide.

65. The 2050 plan summarizes sustainability as "A Hawaiʻi that achieves the following:

- Respects the culture, character, beauty, and history of our state's island communities
- Strikes a balance between economic, social and community, and environmental priorities
- Meets the needs of the present without compromising the ability of future generations to meet their own needs."

66. The Maui Agriculture Design Conference convened a group of approximately fifty whole-systems thinkers and farmers from throughout Maui County along with resource people from Hawaiʻi and the mainland for five days (July 28 through August 1, 2010) in Huelo, Maui, to map out a future for whole systems sustainable agriculture. The retreat was funded in part by the County of Maui.

REFERENCES

Abbott, I. 1992. *Lā'au Hawai'i: Traditional Hawaiian Uses of Plants*. Honolulu, HI: Bishop Museum Press.

Advertiser Publishing Co. 1937. *Hawaii's Crop Parade*. Honolulu, HI: Advertiser Publishing Co. Ltd.

Allen, M. S. 2001. *Gardens of Lono: Archaeological Investigations at the Amy B. H. Greenwell Ethnobotanical Garden, Kealakekua, Hawai'i*. Honolulu, HI: Bishop Museum Press.

Allen, T. 2006. *Tattoo Traditions of Hawai'i*. Honolulu, HI: Mutual Publishing.

Altonn, H. 2004. "Kohala Study Finds Farming Plots Crucial; An Agriculture System Helped to Sustain Kamehameha's Army." *Honolulu Star-Bulletin*, June 13.

Andrade, C. 2008. *Hā'ena: Through the Eyes of the Ancestors*. Honolulu: University of Hawai'i Press.

Andrews, L. [1865] 2003. *A Dictionary of the Hawaiian Language*. With an introduction to the new edition by Noenoe K. Silva and Albert J. Schutz. Honolulu, HI: Island Heritage Publishing.

Ariyoshi, G. 2003. *Diversified Agriculture, Hawaii's Undervalued Resource*. Honolulu: Hawai'i Business Publishing. http://www.allbusiness.com/north-america/united-states-hawaii /693160-1.html.

Athens, J. S. 1997. "Hawaiian Native Lowland Vegetation in Prehistory." In *Historical Ecology in the Pacific Islands*, edited by P. V. Kirch and T. L. Hunt, 248–270. New Haven, CT: Yale University Press.

Beckwith, M. 1970. *Hawaiian Mythology*. Honolulu: University of Hawai'i Press.

Begely, B. W., H. Spielmann, and G. R. Vieth. 1981. *Poi Consumption: Consumption of a Traditional Staple in the Contemporary Era, in Honolulu, Hawaii*. Departmental paper 54. Hawai'i Institute of Tropical Agriculture and Human Resources, University of Hawai'i.

Bishop Museum. 2011. *Good Guys—Bad Guys*. Hawai'i Biological Surveys Program. Honolulu, HI: Bishop Museum. http://hbs.bishopmuseum.org/good-bad/list.html.

Bowers, F. A. I., D. L. Plucknett, and O. R. Younge. 1964. *Specific Gravity Evaluation of Corm Quality in Taro*. Circular no. 61, Hawaii Agriculture Experiment Station. Honolulu: University of Hawai'i.

Buchanan, L., P. Levin, and K. Pali. 2010. *Hui Kala: Acknowledging Hewa; Bringing Everything into Accord—Improving Biodiversity Conservation Efforts through a Deeper Understanding of Cultural Practices and Protocols*. Presented at Hawai'i Conservation Conference, Honolulu, HI, August 4–6.

Burney, D. A., and W. K. P. Kikuchi. 2006. "A Millennium of Human Activity at Makauwahi Cave, Maha'ulepu, Kaua'i." *Human Ecology* 34: 219–247. doi:10.1007/s10745-006 -9015-3.

Campbell, S., and L. K. Mentin, eds. 1986. *Sugar in Hawaii: A Guide to Historical Resources*. Compiled for the Hawai'i State Foundation on Culture and the Arts and the Hawaiian Historical Society, Honolulu, HI.

Collerson, K. D., and M. I. Weisler. 2007. "Stone Adze Compositions and the Extent of Ancient Polynesian Voyaging and Trade." *Science* 317 (5846): 1907–1911.

Cooke, G. P. 1949. *Moolelo O Molokai: A Ranch Story of Molokai.* Honolulu, HI: *Honolulu Star-Bulletin.*

Coral Reef Assessment and Monitoring Program. 2008. "Island: Molokaʻi." http://cramp .wcc.hawaii.edu/Watershed_Files/Molokai/WS_Molokai_molokai_Palaau.htm.

Cottrell, C. A. 2002. *Slivers of Sandalwood, Islands of ʻIliahi: Rethinking Deforestation in Hawaiʻi, 1811–1843.* Master's thesis. Honolulu: University of Hawaiʻi.

Cox, T. R. 1991. *The Birth of Hawaiian Forestry: The Web of Influences.* Paper presented at the 17th Pacific Science Conference, Honolulu, HI, May 27–June 2.

Decision Analysts Hawaiʻi, Inc. 2009. *Maʻalaea Mauka Residential Subdivision: Impact on Agriculture.* Maʻalaea, Maui: Draft Environmental Impact Statement, Ohana Kai Village subdivision.

Diong, C. H. 1982. *Population Biology and Management of the Feral Pig* (Sus scrofa L.) *in Kipahulu Valley, Maui.* Doctoral dissertation. Honolulu: University of Hawaiʻi.

Ellis, W. 1825. *Journal of William Ellis: A Narrative of an 1823 Tour through Hawaiʻi or Owhyhee, with Remarkes on the History, Traditions, Manners, Customs and Language of the Inhabitants of the Sandwich Islands.* Reprint. Honolulu, HI: Mutual Publishing, 2004.

Evans, D., ed. 2008. *Mauka to Makai: Taro Production in Hawaiʻi.* 2nd ed. Honolulu: University of Hawaiʻi Press.

Giffard, W. M. 1913. *Some Observations on Hawaiian Forests and Forest Cover in Their Relation to Water Supply.* Report to the Joint Committee on Forestry, Honolulu, HI.

Gon, S. ʻOhukaniʻohiʻa III. 2010. "Determining the Pre-contact Hawaiian Footprint on Native Ecosystems: Modeling and Traditional Knowledge United." Panel presentation, Agriculture and Hawaiian Ecosystems, Hawaiʻi Conservation Conference, Honolulu, HI, August 4–6.

Hackler, R. 1986. "Alliance or Cessation? Missing Letter from Kamehameha I to King George III of England Casts Light on 1794 Agreement." *The Hawaiian Journal of History* 20: 1–12.

Hall, W. 1904. *The Forests of the Hawaiian Islands.* US Department of Agriculture. Washington, DC: Government Printing Office.

Handy, E. S. Craighill. 1940. *The Hawaiian Planter,* Vol. 1, *His Plants, Methods and Areas of Cultivation.* Bulletin 161. Honolulu, HI: Bernice P. Bishop Museum.

Handy, E. S. C., and E. G. Handy. 1991. *Native Planters in Old Hawaii: Their Life, Lore, and Environment.* With Mary Kawena Pukui. Bulletin 233. Honolulu, HI: Bernice P. Bishop Museum.

Hansford, D. 2007. "Early Polynesians Sailed Thousands of Miles for Trade." *National Geographic News,* September 27. http://news.nationalgeographic.com/news/2007 /09/070927-polynesians-sailors.html.

Hawaiʻi 2050 Sustainability Task Force. 2007. *Hawaiʻi 2050 Sustainability Plan.* Honolulu, HI. http://hawaii2050.org/index.php/site/sp_whatIsSustainability/P1.

Hawaiʻi Department of Agriculture. 2011. *2009 Statistics in Hawaiʻi Agriculture.* Honolulu, HI: Hawaiʻi Department of Agriculture and US Department of Agriculture.

Hawaiʻi Department of Land and Natural Resources. 2007. *Status of Maui's Coral Reefs.* Briefing report produced by the Department of Aquatic Resources and Coral Reef Assessment and Monitoring Program, Honolulu, HI.

Hawai'i Weed Risk Assessment database. 2011. United States Forest Service, Pacific Island Ecosystems at Risk (PIER). http://www.hear.org/pier/wra/pacific/typha_latifolia _htmlwra.htm.

Hollyer, J. R., R. S. de la Pena, K. G. Rohrbach, L. M. LeBeck, eds. 1990. *Taro Industry Analysis No. 4*. Agricultural Industry Analysis: Status, Potential and Problems of Hawaiian Crops series. Submitted to the Governor's Agriculture Coordinating Committee. Honolulu, HI. August 29.

Hutchins, W. A. 1946. *The Hawaiian System of Water Rights*. Honolulu, HI: Board of Water Supply, City and County of Honolulu.

'I'i, J. P. 1959. *Fragments of Hawaiian History*. Translated by Mary Kawena Pukui. Honolulu, HI: Bishop Museum Press.

Iokepa, J. n.d.-a. *Taro Culture*. Collected by Theodore Kelsey. Ms. doc. no. 253. Bishop Museum Archives, Honolulu, HI.

———. n.d.-b. *Hawaiian Taro Planting*. Collected by Theodore Kelsey, Hilo. Hawaiian Ethnographic Notes Vol. 1, 947–949. Bishop Museum Archives, Honolulu, HI.

International Union for Conservation of Nature (IUCN), United Nations Environment Program (UNEP), and World Wildlife Fund (WWF). 1991. *Caring for the Earth: A Strategy for Sustainable Living*. Gland, Switzerland and Earthscan, London: IUCN.

Kagawa, A., and P. Vitousek. 2012. "The *Ahupua'a* of Puanui—a Resource for Understanding Hawaiian Rain-fed Agriculture." *Pacific Science* 66 (22):161–172.

Kalokuokamaile, Z. P. K. 1922. "Ke Ano o ke Kalaiaina." *Nupepa Kuokoa*, May 11 and June 22.

Kamakau, S. K. 1992. *Ruling Chiefs of Hawaii*. rev. ed. Honolulu, HI: Kamehameha Schools Press.

Kelly, M. 1983. *Na Mala o Kona, Gardens of Kona: A History of Land Use in Kona, Hawai'i*. Bishop Museum Report 83-2. Honolulu, HI: Bernice P. Bishop Museum.

———. 1989. "Dynamics of Production Intensification in Pre-Contact Hawaii." In *What's New? A Closer Look at the Process of Innovation*, edited by S. E. van der Leeuw and R. Torence, 82–105. London: Unwin Hyman.

Kepler, A., and F. Rust. 2011. *The World of Bananas in Hawai'i: Then and Now*. Haiku (Maui), HI: Pali-O-Waipi'o Press.

Kirch, P. V. 2011. "When Did the Polynesians Settle Hawaii? A Review of 150 Years of Scholarly Inquiry and a Tentative Answer." *Hawaiian Archaeology* 12: 3–26. http://berkeley .academia.edu/PatrickKirch.

———. 1994. *The Wet and the Dry*. Chicago: University of Chicago.

———. 1985. *Feathered Gods and Fishhooks: An Introduction to Hawaiian Archaeology and Prehistory*. Honolulu: University of Hawai'i Press.

———. 1982. "Transported Landscapes." *Natural History* 91 (Dec.): 32–35.

Kirch, P. V., O. Chadwick, S. Tuljapurkar, T. N. Ladefoged, M. Graves, S. Hotchkiss, and P. Vitousek. 2007. "Human Ecodynamics in the Hawaiian Ecosystem, from 1200–200 BP." In *The Model-Based Archaeology of Socionatural Systems*, edited by T. A. Kohler and S. E. van der Leeuw, 121–139. Santa Fe, NM: School for Advanced Research.

Kirch, P. V., and T. L. Hunt, eds. 1997. *Historical Ecology in the Pacific Islands*. New Haven, CT: Yale University Press.

Ladefoged, T. N., P. V. Kirch, S. M. Gon III, O. A. Chadwick, A. S. Hartshorn, and P. M. Vitousek. 2009. "Opportunities and Constraints for Intensive Agriculture in the Hawaiian Archipelago Prior to European Contact." *Journal of Archaeological Science* 30: 1–10. doi:10.1016/j.jas.2009.06.030.

Laenui, P. 1993. "Indigenous Peoples and Their Relationship to the Environment." Collected by Pōkā Laenui and presented to the Multi-Cultural Center at California State University Sacramento. http://hawaiianperspectives.org/environ.txt.

Lebo, S. A., J. E. Dockell, and D. Olszewski. 1999. *Life in Waipio Valley, Hawaii, 1880–1940.* Honolulu, HI: Bishop Museum.

Levin, P. 2006. *Statewide Strategic Control Plan for Apple Snail,* Pomacea canaliculata, *in Hawai'i.* Report produced for the Department of Aquatic Resources, Department of Land and Natural Resources, Honolulu, HI.

Liliuokalani of Hawaii (Queen Lili'uokalani). 1897. *The Kumulipo: An Hawaiian Creation Chant. An Account of the Creation of the World According to Hawaiian Tradition.* Reprint. Honolulu, HI: Pueo Press, 1978.

Lukens, A. 2010. "Intersections: The Story of Maui's Four Great Waters." *Green Magazine* 2 (2).

MacCaughey, V., and J. S. Emerson. 1913–1914. "The Kalo of Hawaii." *Hawaiian Forester and Agriculturist (Honolulu, HI)* 10 (7–12) and 11 (1, 4, and 7).

Malo, D. 1898. *Moolelo Hawaii: Hawaiian Antiquities.* 2nd ed. Translated by N. B. Emerson. Honolulu, HI: Bishop Museum Press.

Maly, K., and O. Maly. 2005. *He Mo'olelo 'Āina No Ka'eo Me Kāhi 'Āina E A'e Ma Honua'ula O Maui: A Cultural-Historical Study of Kaeo and Other Lands in Honuaula, Island of Maui (TMK 2-1-07:67).* Prepared for Sam and Jon Garcia by Kumu Pono Associates, Lana'i City, HI.

Merlin, M., and D. VanRavenswaay. 1990. *The History of Human Impact on the Genus* Santalum *in Hawaii.* Presented at the Symposium on Sandalwood in the Pacific, Honolulu, HI, April 9–11.

Moir, W. W. G. 1933. *The Native Hawaiian Canes.* Bulletin 7: Proceedings of the Fourth Congress of the International Society of Sugar Cane Technologist, San Juan, Puerto Rico.

"Molokai: Future of a Hawaiian Island." 2008. Prepared by members of the Molokai Community. Kaunakakai, HI. http://sustainablemolokai.org/16-2/.

Mookini, E. K. 1974. *The Hawaiian Newspapers.* Honolulu, HI: Topgallant Publishing Company, Ltd.

Morton, J. F. 1987. "Breadfruit." In *Fruits of Warm Climates.* Miami, FL. http://www.hort.purdue.edu/newcrop/morton/index.html.

National Agriculture Statistics Service. 1946–2013. "Hawaii [various statistics]." http://www.nass.usda.gov/Statistics_by_State/Hawaii/Search/index.asp.

Natural Step. 2009. *Hawai'i Sustainability Primer: Step by Natural Step.* Ottawa, Ontario: The Natural Step. http://www.kohalacenter.org/pdf/PrimerGuidebookHawaiiFINAL.pdf.

Nogelmeier, M. P. 2010. *Mai Pa'a I Ka Leo: Historical Voice in Hawaiian Primary Materials, Looking Forward and Listening Back.* Honolulu, HI: Bishop Museum Press and Awaiaulu.

Olszewski, D., ed. 2000. *The Māhele and Later in Waipiʻo Valley, Hawaiʻi.* Honolulu, HI: Bishop Museum.

ʻOnipaʻa Nā Hui Kalo. 2003. *Guidelines for Grassroots Loʻi Kalo Rehabilitation: Pono, Practical Procedures for Loʻi Kalo Restoration.* Printed with the support of Queen Liliʻuokalani Children's Center Windward Unit. Punaluʻu, HI: ʻOnipaʻa Nā Hui Kalo.

Ooka, J., and B. Brennan. 2000. *Crop Profile for Taro in Hawaii.* Honolulu: University of Hawaiʻi.

Page, C., L. Bony, and L. Schewel. 2007. *Island of Hawaii Whole System Project: Phase I Report.* Rocky Mountain Institute on behalf of the Omidyar family, Hawaiʻi.

Preza, D. 2010. *The Empirical Writes Back: Re-Examining Hawaiian Dispossession Resulting from the Mahele of 1948.* Masters thesis. Honolulu: University of Hawaiʻi.

Pukui, M. K., and S. H. Elbert. 1986. *Hawaiian Dictionary, Revised and Enlarged Edition.* Honolulu: University of Hawaiʻi Press.

Queen Emma (Emma Kaleleonalani Naʻea Rooke Kamehameha). n.d. [c. mid-1800s]. Observations on Varieties and Culture of Taro. Unpublished manuscript. Queen Emma collection 71, Bishop Museum Archives. Honolulu, HI.

Reeve, R. n.d. *Na Wahi Pana O Kahoʻolawe: The Storied Places of Kahoʻolawe: A Study of Traditional Cultural Places on the Island of Kahoʻolawe.* Consultant report no. 17, prepared for the Kahoʻolawe Island Conveyance Commission, Honolulu, HI.

Reppun, C. 2010. "Agriculture." In *The Value of Hawaiʻi: Knowing the Past, Shaping the Future,* edited by Craig Howes and Jon K. Osorio, 39–46. Honolulu: Biographical Research Center, University of Hawaiʻi Press.

Salmoiraghi, F., and Y. Yoshinaga. 1974. *Waipio: An Exhibition at the Wailoa Center.* Exhibition catalog. Hilo, HI: Wailoa Center.

Schenk, S., M. W. Crepeau, K. K. Wu, P. H. Moore, Q. Yu, and R. Ming. 2004. "Genetic Diversity and Relationships in Native Hawaiian *Sacchinum officinarum* Sugarcane." *Journal of Heredity* 95 (4): 327–331.

Sedgwick, T. F. 1902. *The Root Rot of Taro.* Bulletin no. 2, Hawaii Agricultural Experiment Station, Honolulu. Honolulu: Hawaiian Gazette Company.

Smith, J. G. 1902. *Annual Report of the Hawaii Agricultural Experiment Station, 1901 and 1902.* Washington, DC: US Department of Agriculture, Office of Experiment Stations.

Smith, J., and S. A. El-Swaify. 2006. *Towards Sustainable Agriculture: A Guide for Hawaiʻi's Farmers.* Honolulu: University of Hawaiʻi.

Social Science Research Institute. 1998. *Hoʻōla Hou I Ke Kino o Kanaloa: Kahoʻolawe Environmental Restoration Plan.* Prepared for the Kahoʻolawe Island Reserve Commission. Honolulu: University of Hawaiʻi.

Sohmer, S., and R. Gustafson. 1987. *Plants and Flowers of Hawaiʻi.* Honolulu: University of Hawaiʻi Press.

Stannard, D. E. 1989. *Before the Horror: The Population of Hawaiʻi on the Eve of Western Contact.* Honolulu: University of Hawaiʻi.

Sterling, E. 1998. *Sites of Maui.* Honolulu, HI: Bishop Museum Press.

Taro Security and Purity Task Force. 2009. *E ola hou ke kalo; hoʻi hou ka ʻāina leiʻa. The taro lives; abundance returns to the land.* Taro Security and Purity Task Force 2010 Legislative Report. Honolulu, HI: Office of Hawaiian Affairs. http://www.hawaii kalo.org.

Vitousek, P. M., T. N. Ladefoged, P. V. Kirch, A. S. Hartshorn, M. W. Graves, S. C. Hotchkiss, S. Tuljapurkar, and O. A. Chadwick. 2004. "Soils, Agriculture and Society in Precontact Hawai'i." *Science* 304 (June): 1665–1669.

Wagner, W. L., D. R. Herbst, and S. H. Sohmer. 1999. *Manual of Flowering Plants of Hawai'i.* Honolulu: University of Hawai'i Press and Bishop Museum Press.

Whelan, J. A., and L. K. Menton. 1988. *Ranching in Hawaii: A Guide to Historical Resources.* Honolulu: State Foundation on Culture and the Arts and Hawaiian Historical Society.

White, L. D. 1994. *Canoe Plants of Ancient Hawaii.* Waitsfield, VT.

Whitney, L. D., F. A. I. Bowers, and M. Takahashi. 1939. *Bulletin 84: Taro Varieties in Hawaii.* Honolulu: Hawaii Agriculture Experiment Station, University of Hawai'i.

Wilcox, C. 1996. *Sugar Water: Hawaii's Plantation Ditches.* Honolulu: University of Hawai'i Press.

Wilder, G. P. n.d. "The Kalo in Hawaii." MsSc, Wilder Box 24. Bishop Museum Archives, Honolulu, HI.

World Commission on Environment and Development. 1987. *Our Common Future* (The Brundtland Commission Report). Oxford: Oxford University Press.

Zerega, N., D. Ragone, and T. Motley. 2004. "Complex Origins of Breadfruit (*Artocarpus altilis*, Moraceae): Implications for Human Migrations in Oceania." *American Journal of Botany* 9 (15): 760–766.

Zschokkee, T. C. 1931. *The Problem of Soil Saving in the Hawaiian Islands.* Extension Bulletin no. 11. Honolulu: Agriculture Extension Service, University of Hawai'i.

FOUR

Lessons from the Taro Patch

PENNY LEVIN

Aia i ka mole ke ola; e 'ike pono i ke au nui me ke au 'iki
There in the foundation is life; know well the big currents
    and the little currents

The revitalization of traditional agriculture, particularly *kalo* (taro) farming, in Hawai'i is viewed by some as a step back in time and by others as a return to center, especially when it comes to food self-sufficiency. Chapter 3 of this volume makes clear that *kalo* farming was serious agriculture that rivaled the sugar industry and put food directly into the calabash[1] of hundreds of thousands of people. In the growing of *kalo*, Hawaiians also retained a strong understanding of the big and small currents that affected the sustainability of food production in the islands. This chapter examines the *mole* (foundation) and the *hi'ohi'ona* (elements) that form the traditions of growing *kalo*; the first offers key concepts, and the latter considers the details and provides examples of how that manifests in farm practice.

It is important to stress that this is by no means a complete discussion. There are many ways to express or expand on each idea presented here. Indeed, it is possible to sum up the entirety of this section with the Hawaiian words *pono* ("to be in balance" is but a limited definition)[2] and *ho'oponopono* (to make right), for which much has been written (e.g., Chun 2011, 1–13; Wianecki and Kanahele 2011; Pukui, Haertig, and Lee 1983). For such a general statement to work, however, requires a shared understanding of what that might mean on the farm, in a community, and at the policy level. This is what allowed Hawaiians to create the natural resource management systems they did and, reciprocally, for those systems to function well, something that does not often exist today for a variety of reasons. It is the intention of this chapter to invite thoughtful exchange. The Hawaiian words in this chapter are a window into a richer dialogue that emerges from the taro patch. I wish to thank the many taro planters who provided input and review to this chapter.

Lest the reader think this is solely a discussion of "what was," it is important to understand the staunch survival of *kalo* farming despite its many challenges, and the revival of taro cultivation that has slowly unfolded since the 1970s. That time frame marks the beginnings of the largest legal cases in Hawai'i calling for the return of water from plantation ditch systems back to the taro-growing lands and streams they originated from[3] and coincided with a revival of Hawaiian culture. This period has also seen the emergence of new grassroots taro-growing organizations statewide that recognize the importance of sharing their experiences with each other, helping families and communities re-open old *lo'i* lands (wetland taro fields), and helping to "grow" more taro farmers. Since 2008, despite the passing of old-time planters, there are new taro farmers, both subsistence and commercial, and more young adults engaged in *kalo* cultivation than in the last decade.[4] Gardens are coming alive again in schools throughout the state, and taro is a part of the program (see, e.g., chapter 11 of this volume).[5] There is a resurgence in taro cultivar diversity.[6]

While commercial production of raw taro and milled poi has remained somewhat level over the last decade,[7] it is among the small growers and the return to hand-pounded *pa'i 'ai* and poi that change is happening. Practitioners in the forefront of this effort estimate as many as 8,400 people, young and old, have learned how to make *kalo* into *pa'i 'ai* and poi using a traditional board and stone in classes and at events over the last five years, and that number is growing rapidly. More than 230 boards and 340 stones have been distributed to schools, communities, and families in that same time in support of revitalizing the tradition of food self-sufficiency.[8] Families are dusting off the boards and stones of their *kūpuna* (grandparents, ancestors) and bringing them back to life.

Although statistics are not available at this time, a small number of young farmers on each island are stepping up to grow taro commercially, something not seen since the 1970s and 1980s. Families are growing their own *kalo*, buying from farmers' markets, or purchasing directly from the farm to make food for their own tables. In 2011, the State of Hawai'i (re)legalized the making and selling of hand-pounded *kalo* direct to consumers. This is broadening consumer palettes, creating higher-value niche markets for smaller commercial growers, and creating income-generating opportunities for young *ku'i 'ai* (poi pounding) practitioners.[9] Thanks to a small but expanding group of dedicated farmers and practitioners, the sound of the poi pounder is no longer an echo of the past but a *kani* (a reverberating beat) for the future.

The revival of hand-pounded poi is just beginning to have ripple effects on the farm. Commercial growers typically know they can push the taro for size and that a percentage of *loli* (overripe; watery) taro ends up in the harvest for the poi mills, whose machinery are somewhat forgiving in the blending of thousands of pounds of taro. Hand pounding requires a higher-quality, dense, starchy corm to produce good poi. To achieve that, growers who wish to supply this new market are having to reevaluate conventional taro production methods, soil management strategies, taro varieties, and the timing of harvests. In commercial and subsistence sectors, depleted soils and high rates of soft rots are also challenging taro farmers to search for alternative soil and crop management practices.

LESSONS IN SUSTAINABILITY FROM THE TARO PATCH

Taro is sibling, sacred offering, food source, medicine, a source of dyes, a trade commodity (from its earliest exchanges of fish for poi), and a modern market crop. In reality, we do not know how sustainable Hawaiian agricultural systems were. At the peak of the population, there was tremendous competition for organic matter to feed the soil, particularly in rainfed regions. Based on the evidence presented in chapter 3 of this volume, it would be a disservice to today's taro growers and to *'āina* (the land; see page 87 for a more in-depth discussion on the meaning of *'āina*) to assume that a static or perfect world of sustainable taro practice existed. More likely it was always in flux, either moving away from or into balance,[10] depending on climatic and sociopolitical conditions, as well as the state of surrounding natural resources. The key, however, was a common understanding of where "center" was: that sweet spot of soil, crop, and *ahupua'a* (traditional land division) health and how things connected. The challenge is to discern the details of restorative practice in taro farming from the written and ethnographic accounts of the past, through *kūpuna* (elders) still living, and from observations in the present. We have more available to us to learn from than we imagine.

The eloquence of the Hawaiian language, born from the islands and the sea for which it was created, gives rise to words that are rich with multiple meanings. *Mole* (pronounced mo-leh) is both a foundation and a source—a place

Do we know where "center" is?

Mole = foundation and source

of beginnings and underpinnings. Just as a foundation is a base that is often unseen, so too, the *mole*. The word is inhabited by the *kalo*, first as taproot and ancestral root and as the name given to the smooth, round bottom of a *pōhaku ku'i 'ai* (stone pounder), the tool traditionally used to turn *kalo* into poi (Andrews [1865] 2003; Pukui and Elbert 1986). The link between the *mole* of the stone poi pounder and the *kalo* as a cultural foundation is not by accident. In the Hawaiian cosmology, man descended from the *kalo*. The role of *pōhaku* in ancient Hawai'i ranged from the sacred, storied, and foundational to the tools of everyday life. To the *mahi'ai* (farmer), the stones furnished the means to build their terraces and waterways and to form implements that enabled him or her to grow and make food, to fish, to build a canoe or a house.[11] While modern technology has replaced some of these tools, the *pōhaku ku'i 'ai* and the stone terraces of the *lo'i kalo* (wetland taro fields) remain an essential part of taro farming and food production. Five *mole* are considered here: *Relationship, Agriculture fit to the land, Resilience, Locally sustainable and renewable,* and *Leads to abundance.*

Each of these *mole* is further broken down into *hi'ohi'ona,* or its visible, physical elements and characteristics. Andrews ([1865] 2003) relates *hi'ohi'ona* to the features of a person: his peculiar gait and visage, the form or external appearance of something. Pukui and Elbert (1986) describe it also as the features of a landscape: the topography and the plants, rocks, soil, water, and resources located there—the very things that influenced the details of traditional agricultural design, size, and practice.

Mole: *Relationship. Ho'i hou i ka mole* (Return to the taproot)
(*'Ōlelo No'eau* no. 1025)

Why start here? *Kūpuna* (elders) encourage us always to look to the source and begin with the very roots of agriculture, with recognition of the relationships that arise from the growing of *kalo*. Of all of the differences between conventional agriculture and *kalo* production, perhaps this is the one thing that stands out, albeit quietly, the most.

For some taro farmers, there is a kin relationship to Hāloa, the plant; for others, it is an ecosystem perspective, the day-to-day recognition that the *lo'i* depends on and is connected to everything *mauka* and *makai* (from the uplands to the reefs); that this crop demands attention to the whole of the *ahupua'a* (traditional land division). For still others it holds their families together or is a connection between community and market. Growing *kalo* is backbreaking work, and still it is a crop that invites marvel and affection, contemplation, and care, that responds to *alo-ha:* love-shared breath.

Do farmers talk about the love they have for their crops? Indeed, but mostly they just tend to business.

Found within the *mole* of *Relationship* is the **Hiʻohiʻona** Sacred and practical

A common denominator in indigenous agriculture throughout the world is the recognition of primary staple crops (and the earth as a whole) as sacred elements, gifts from or embodiments of the gods, and often parent (or older sibling) to man. Across Asia, rice was honored as a living being. In North America, the Ojibwe honored Manoomin, the wild rice, as a gift from the gods. The connection of older sibling to younger sibling is reflected in a Haudenosaunee (Iroquois) creation myth that tells how the daughter of Sky Woman, pregnant with twin boys, died in childbirth. From the place she was buried grew three plants—corn, beans, and squash—that fed her sons and all of her descendants.[12] So it was in South America where the Maya also recognized that man was born of the corn. And in turn, the stillborn child of Hoʻohokukalani, daughter of Papa and Wākea, was buried in the ground and the *kalo*, Hāloa, sprouted forth.

Mircea Eliade's famous work *The Sacred and the Profane* (1959) acknowledged that "nature is never only 'natural'; it is always fraught with a religious value."[13] Yet, the western mind struggles with this idea; there appears to be little room in modern farming for such concepts. The sacred is kept separate from science, the practical from the cultural.

The biologist and physicist Lyall Watson "recognized the need for reconciliation between man and nature" and wrote, "There seem always to have been two ways of looking at the world. One is the everyday way in which objects and events, although they may be related causally and influence each other, are seen to be separate. And the other is a rather special way in which every thing is considered to be part of a much greater pattern." Watson alleged that "we didn't come into this world. We came out of it, like buds out of branches . . . and if we turn out to be intelligent beings, then it can only be because we are fruits of an intelligent earth" (1991, 36–37).

One need only look into the cultural rituals of traditional farming to see that behind the sacred is the practical, as well as a solid track record of indigenous science. The role of the practical in sustainable agriculture is easily discerned, but to what purpose the sacred?

The presence of the sacred causes us to pause and choose wisely, to pay attention and work with respect. It reminds us of the interconnections between the heavens, water, soil, and self, and our dependence on all that feeds us. It requires us to be more careful in the daily practice of tending the *kalo,* in the harvest, in the making and eating of poi. It engenders appreciation for the food on our plates and keeps us from taking it all for granted. It is what distinguishes food from commodity and provides nourishment for the soul along with the body. Embracing the sacred shifts us from stewardship to kinship (Laenui 1993).

Behind ritual and *kapu* (symbolic codes of the sacred) are well-honed natural resource management and community relationship strategies. Perhaps it is as simple as the understanding that in order "to make sense, you must have sensed" (Watson 1991, 30). The senses are the foundation tools of all observations *and* the core of good science, indigenous or otherwise—the most important tool a farmer has.[14] That includes "gut instinct." To understand how the cycles of the moon affect planting, to know in the rainy season that a year of drought is coming, to know from the *ao* (the clouds)[15] that a storm will appear in three days, or to know through the feet that the mud in the *lo'i* is not right comes from years of observation. From this evolves *'ike pāpālua,* deep insight: the point where the imagined boundaries between culture and science, the practical and the sacred disappear.

Retaining the connection of older brother–younger brother between *kalo* and Hawaiians created on a very practical level the understanding that if the *kalo,* and the soil and water that fed it, were well tended, the population would not go hungry. Equally so, a farmer would strive to grow the healthiest and highest yields just as he would strive to give the best care to a *punahele* (favorite child).

There is much in the history of Hawai'i that has worn away at the *'aha 'ula* (sacred cord) between man and the land. If sustainable agriculture is to be more than a passing trend and shift what we choose to grow in Hawai'i and how we choose to grow it, it must be more than just economically efficient or ecologically sound. It is a *na'au* (heart/gut) relationship that will help us stay the course.

A second aspect of *Relationship* is the **Hi'ohi'ona,** *Kū and Hina* (male and female)

The pairing of opposites permeates Hawaiian language and storytelling. The *Kumulipo,* one of the oldest and longest genealogical chants, unfolds as a long line of pairs of organisms that populate the islands, beginning with the night (*pō*) and the day (*ao*) and proceeding through the birth of all the things of the sea and their *kia'i* (guardians) on the land. The narrow waters (male) and the broad waters (female) of the refrain bind the process of creation and evolution together. Among the Hawaiian gods, Kū, the male, and Hina, the female, are key elements in the reproductive energy of life.[16] They do not just symbolize a pairing of opposites but also balance each other in nature. Without this, the plants of the uplands, the crops in the field, and the things of the streams and the ocean do not bear fruit.

For the farmer, awareness of opposites in pollination, plant forms, weather patterns, landscape features, days and seasons, and even field preparation and mulching traditions influenced best management practices and increased planting success. Kū and Hina names and their associated plant shapes might indicate to a farmer that strong winds, lack of water, oversaturation, or deep or poor soil substrate had to be considered in the location or design of plantings.[17] The results of *kū* (upright) and *hina* (prostrate) methods of planting *kalo* are different. One yields more *keiki* (children; shoots or suckers) than the other, an important thing to know if the objective is to create more planting material. Kū and *papa* or *komo* (prostrate, flat or crawling) plant forms (*papa* being associated with Hina) were found among food cultivars. A taro plant that stood *kū* (tall and straight), such as a Lauloa variety, would overshadow shorter, stockier types if planted in the same field. The famous *kō 'eli lima* (the "hand-dug" sugarcane) and the *'ulu hua i ka hāpapa* ("crawling" breadfruit) of Ni'ihau were buried in the sand with only their leaves showing above the dunes. An *'ulu papa*, a breadfruit tree with low stature and downward branches, was easier to harvest than an *'ulu kū* (a tall-standing breadfruit tree).[18]

Recognizing the male and female characteristics and roles in *pōhaku*, plants, and planting practices celebrated the fecundity of life. Samuel Kamakau (1976, 34) describes the softening of the mud in a new taro patch as "a great day for the men, women, and children, and no chief or chiefess held himself too tabu to tread in the patch. Every man, woman, and child bedecked himself with greenery, and worked with all his might—trampling here and there, stirring the mud with his feet, dancing, rejoicing, shouting, panting, and making sport." The teasing banter and play of men and women during work "encouraged" the productivity of the *lo'i* and made life *'ono* (delicious).

Beckwith (1951, 14) and Johnson (1981) also suggest that a multitude of contrasting elements shaped and represented economic exchange and resource distribution and management in ancient Hawai'i, particularly in the sharing of resources *mauka* to *makai*.

Hiʻohiʻona: Humbleness and strength

In the earliest centuries of Polynesian colonization in the Hawaiian Islands, the sibling status between *kalo* and man reflects a practical choice for collective rather than individual survival and rivalry. It bound the farmer to care for this important food plant, ensuring continuity. A well-known Hawaiian proverb, *he aliʻi ka ʻāina; he kauwā ke kanaka* (the land is a chief; man is its servant),[19] makes a fundamental statement about Hawaiian relationship to the land. Tending the land (service) requires humility and great strength. The first keeps us attentive to cause and effect in the environment and turns conventional decision making on its head. The latter is necessary for the hard work and the strength of character needed by taro farmers to protect their livelihoods. The ancient story *Kalo Paʻa o Waiāhole* (Hard Taro of Waiāhole) has come to represent in modern times not so much a stubbornness as steadfastness in the face of great odds as taro farmers fight for the return of water to the streams that give *kalo* life.[20]

To the ancient Hawaiian farmer, *ʻāina* (and Hāloa) sat at the center of all things, not man.[21] Reflecting on recorded chants and the practices of old, a pattern of stating intentions and asking for support from the land and associated gods emerges (e.g., Gutmanis 1983). For the *makaʻāinana* and the *aliʻi* (commoners and chiefs), "What's in it for me?" was realized through a broader lens of "If I care for the land and the *kalo,* there will be enough for those who depend on me." This seems to run counter to mainstream business models in agriculture, which strive to put each producer on top, pitting one grower against another (competition). Agencies and economic analysts reward the largest growers at the expense of the small. Conventional agriculture practices, too, place man above nature and wrestle the land into productivity. While both models have a goal of "profit" and high yields, "profit" in agriculture today is primarily defined as net income, whereas, profit from the perspective of the *loʻi* included the abundance of food, as well as the relationships formed.

Is it naïve or "off center" in today's world to be in a humble relationship with the land? To ask consent of the *ʻāina* before beginning intended work? Looking around at current conditions in the *loʻi* and the state of our natural resources, one could argue that poor conditions exist precisely because we have lost our grasp of this relationship. Hawaiian stories that illustrate food production competition between the chiefs (e.g., as in Keʻanae, Maui) are recorded, but more common are the stories of rewards (from the gods and the people) for good husbandry of the land and the sea. Hoarding, greed, boasting, or poor agricultural practices were repaid with harsh consequences, including lost harvests, the disappearance of resources such as water or schools of fish, and even death.[22]

There is no arguing that a farmer needs to make a dignified income to be able to live and care for his or her family. It is also true that the demands of the market, the small size and income of most family farms in Hawai'i, and a limited labor pool have driven taro farming to a different course. Is there room, then, for entrepreneurship and economic well-being within a tradition of service to the land? The scientist and Buddhist monk Matthieu Ricard wrote, "Humility does not mean believing oneself to be inferior, but to be freed from self-importance" (qtd. in Föllmi and Föllmi 2003). Such a flexible position supports a better rapport with the land, which increases the likelihood of producing a high-quality crop that, in turn, attracts a more committed set of buyers and a better price directly to the farmer. It revalues farmers, not just as food producers but as caretakers.

When we choose our own wants and needs before those of the 'āina, we lose—soil and soil productivity, water resources and cycles, and dependable productivity all decline. As farmers and consumers, we cannot afford to remain in that position. When 'āina comes before self, when it sits in the center of decision making, we open the way for sustainable practice.

Hi'ohi'ona: Interconnected and reconnective

In the Hawaiian language the word 'āina embodies an elegant worldview. 'Āina is both the food and the land that sustains us. Unlike the relatively concrete western concept of land, there is no GPS point that locates where exactly 'āina is. It encompasses the deep ocean, the nearshore reefs, terra firma, and the heavens—the soil, the rocks, and everything that grows upon it, the fresh waters that flow above and below ground, the peak of the mountains, the ao, and the life-giving rains that bridge all these things that we often see as separate. All of the energy moves within and as part of a whole system.

If the land dictates our actions, then what follows is an understanding of one of the core principles of ecology and Hawaiian thought—e pili ana āpau, everything is connected.[23] In conventional agriculture practice we have gotten very far away from that. But without it, "sustainable" agriculture does not exist. Observation and discernment affirm this perspective, guide decisions, and develop agricultural strategies that preserve, restore, and maintain critical environmental linkages that support us, particularly soil-building and water-cycling processes. This is something that taro farmers did on a daily basis.

One of the key connective roles that the *lo'i* plays is to support the movement of water between *mauka* (upland) streams and the ocean for freshwater biota. By design, they traditionally ameliorated sediments by capturing silt before it filled in fishponds at the coast. The nutrients and organic matter in turn fed the *kalo*. The oldest systems still preserve designs that slow flood waters so effectively that they do no damage to the fields or the crop ('Onipa'a Nā Hui Kalo 2003).[24] In the lowlands, the fields frequently replaced marshlands but retained the same ecological roles. The string of *lo'i* systems along island coastlines provided a wealth of symbiotic habitat for birds and fish, some of which served as food.[25]

Dryland *kuaiwi* (rock wall) systems worked in similar fashion. Stone walls in upper-elevation systems captured and controlled downhill flows of organic matter and soil. The rocks heated during the day and released moisture through condensation during the evening, which watered the plants that in turn attracted additional moisture from the clouds.[26] Dead plant material became the sponge within the rock walls for retaining water and making new soil. Visible within these designs is the intentional maintenance of natural water and nutrient cycles.

Without surface water flows in the streams and ancient *'auwai* (traditional irrigation canals), without *mauka* (upland) forests reaching the lower elevations so that the rain could *hahai* (follow) down the mountains and percolate into basal aquifers,[27] the entire system begins to break down. US Geological Survey reports and the state's Division of Aquatic Resources have documented impacts to every part of the water cycle and the food chain, from the *'ōpae kolo* (mountain shrimp) and the *pinao 'ula* (damselflies) in the upper reaches to the *'o'opu* (goby) and *hīhīwai* and *hapawai* (freshwater shellfish) that move between the mountains and the ocean, to the *kalo* in the fields, the *āholehole* and *'āweoweo* (fishes) between to the whole of the coral reefs and the health of nearshore and deep-sea fisheries (Oki, Wolff, and Perreault 2010; Parham et al. 2008; Gingerich et al. 2007). Today, taro farms offer sometimes the only connection where fresh water can still reach the ocean, if only because their presence prevents the wholesale extraction of water from a stream or *ahupua'a*. Rehabilitating old *lo'i* sites is a practical, multibenefit model for restoring natural cycles district by district.

Interconnection and reconnection are as much about community as the environment. In traditional culture, the two are hard to separate. A primary thread that binds each family in a taro-farming community, and the farms themselves to the *ahupua'a*, is water and the main *'auwai* (irrigation canal) that carries water through a field system. The *kalo*, too, provided a bridge that connected people to each other. One outcome of such relationships in Hawai'i was collaborative and reciprocal labor and care of resources.

Collective preparation of fields, planting and harvest, fishing, and building are described in well-known literature (Kamakau 1976; Fornander 1916–1919; Malo 1898). *Kō'ele* and *lo'i pō'alima* ("Friday patches" named after the workday associated with them) were fields designated for required community labor (a form of tax) for the chiefs. There are several ways in which traditions of shared work continue in the present day and places where they have fallen away.

In commercial wetland and dryland *kalo* production, the use of machinery is limited to tilling fields and mowing or weedeating banks and field edges. For the most part, the rest of the planting-to-harvest cycle remains dependent on manual labor. Without the work of family members, hired labor must fill the gap at significant cost. Farmers with tractors sometimes assist other growers, for a fee or for free, in opening fields and in repair of ditches, particularly after big storms.

Ancient *'auwai* systems, which often start in the mountains in rough terrain, need cooperative tending from the intake in the stream to the top farm in the complex and along the length of the parent canal as it passes each farm and rejoins the stream and eventually the ocean. Up through the 1980s, the main *'auwai* in multifamily *lo'i* systems were cleared regularly by entire communities. The first priority was to keep the river intake and *'auwai* free of vegetation, debris, and stones, the banks stable, and the exit at the ocean clear. Work was done when it was needed, and everyone came. Ditch-clearing days were also a social event where people sometimes made *imu* (an underground oven), sat down, and shared food. This collaboration is one of our biggest losses in taro-farming communities where the practice no longer functions.

When a system is neglected, it begins to fall apart. Today the work is often the burden of one or two farmers at the top of the *'auwai*, many of them aging with no younger growers learning how the entire system works to ensure its proper care in the future. Most taro districts are also hampered by fewer farmers as a whole than in the past. Places such as Waimea, Kaua'i, have such long canals that they become impossible to care for given the existing number of growers.[28] As the costs of maintaining a farm increase, we would do well to rekindle our relationships and *kōkua* (assist) each other so the production of food has a better chance at success.

The revival of community *lo'i* in the 1980s created the opportunity for residents and students outside of existing taro farming families to reconnect to growing *kalo* and sharing in the rewards. Two of the more well-known sites where this occurred are Ka'ala Farm, Wai'anae, and what became known as the Mauka Lo'i in Waiāhole, O'ahu. One of the primary lessons, beyond growing *kalo,* that came out of these projects was shared *kuleana* (responsibility). 'Onipa'a Nā Hui Kalo, a statewide organization of experienced and new taro growers and supporters, has brought their *kōkua* (help) island to island for the last eighteen years, with the assistance of the Queen Lili'uokalani Children's Center, in order to reopen ancient taro lands and make them operational again. Under 'Onipa'a Nā Hui Kalo projects, the land is cleared by hand, from the removal of trees and tall grass to rebuilding stone walls, *lo'i,* and *'auwai.* The sheer number of people working the land and turning the mud together with sweat and laughter results in planted taro patches out of the overgrowth within the span of a single day. Watching this group, it is easy to imagine Kamakau's (1976) description of Hawaiians in the *lo'i* long ago (see p. 85). It brings true meaning to the saying *'a'ohe hana nui ke alu 'ia,* no task is too big when done together by all (Pukui 1983). Community workdays and service learning projects at educational, subsistence, and small-scale commercial taro farms across the state are a reflection of this tradition that give some growers an extra boost in farm maintenance and production and teach participants the power of cooperative effort.

On the farm, the effects of water management from one patch to the next and the amounts, timing, and kinds of inputs each grower uses fall to the next receiver of the water. Every taro farmer is aware or learns quickly that the quality and quantity of the stream water he receives directly impacts crop yields. Today's reduced flows have resulted in daily temperatures higher than 90°F in the *lo'i,* which weakens young plants and promotes the growth of pathogens that attack adult corms.[29] The apple snail, taro's most aggressive pest, proliferates at greater rates in warmer waters and spreads through the liquid connection between patches, making it extremely difficult to eradicate without control measures in sync up and down the system. The warm water, along with polluted runoff from adjacent residential and agricultural lands in some cases, has also increased the presence of *Staphylococcus, E. coli,* and other bacteria in streams and on farms to levels that endanger the health of taro farmers.[30] Reconnecting this single element of water resolves temperature and subsequently many associated issues. The return of water to Waiāhole Stream on O'ahu proved that it can rejuvenate native stream life while helping to flush out competing alien fauna.[31] This is a clear indicator that to achieve sustainable agriculture in Hawai'i requires participation from a broader community than just farmers. The descriptions above reaffirm that what happens in one sector of the web impacts the rest.

Out of the preceding elements emerges a gathering **Hi'ohi'ona:** Whole systems thinking, decisions, and actions

An archipelago or an island resonates a signature across the ocean currents. Navigators used this to find their way across the Pacific to Hawai'i. Early inhabitants observed how geologic and environmental conditions on one island were related to conditions on another. The veracity of this statement is found in accounts of Pele's traverse across the island chain in ancient times and in changing weather patterns between Maui and Kaho'olawe islands following deforestation on the upper slopes of the two facing districts recorded at the turn of the last century (Spoehr 1992; Tenney 1909). Hawaiians demonstrated a keen ability to think in whole systems that culminated with the concept of *ahupua'a*.

The *ahupua'a*, a traditional unit for natural resource management in Hawai'i, is thought to have formalized in the time of Umi-a-Liloa, a chief of Hawai'i Island and Mā'ilikūkahi between the sixteenth and seventeenth centuries (Preza 2010; Kamakau 1992; Fornander 1969). Recognized as both political and ecological boundaries, *ahupua'a* have been defined by inclusion of a minimum set of assets necessary to live, even where no access to the ocean or the highest slopes occurred. From available maps and records, it appears that these land divisions were drawn to enfold the moisture-capturing places,[32] watercourses and cycles within their limits, as well as places to support wetland and dryland agriculture, or both. A series of "habitats" along a continuum created an integrated framework for management—beginning in the *lani* (heavens),[33] descending into the *kuahiwi* (mountain), to the *nahele* (mid- to lowland forests where everyday gathering occurred), the *kula* (the open plain),[34] the *kahakai* (shoreline), the *po'ina* (place where the sea crests and breaks), *kai kohola* (the shallow sea within the reefs), and ending in the *kai uli* (the deep blue ocean).[35] Overarching was the *wao akua* (region of the gods) and the *wao kanaka* (the region of men) defining what and where activities occurred.

Conventional descriptions of how the *ahupua'a* worked are often oversimplified and brief. They focus on the chiefs and the *konohiki* who managed the lands of a chief, his authority over the people and produce of the lands, and his responsibilities to collect and transfer those goods to his *ali'i* and the gods, and suggest each *ahupua'a* had a complete set of resources. Preza (2010, 61–64) refutes the generalizations of the "pie shaped" *ahupua'a* for its misrepresentation of resource inclusion and allocation.[36] Malo (1898, 59) interestingly writes "those who farmed the land of the king or chief were *konohiki*," suggesting they were active participants and not just overseers in the work of the land. The wisdom to administer the entirety of ecological and human intersections within an *ahupua'a* for planned long-term resource availability stemmed from several sources, including training, regular consultation with *kūpuna* (elders), *kahuna* (specialists), and skilled *maka'āinana* (people who tended the land) who were well versed in customary knowledge, the landscape around them, natural and spiritual phenomena, and the carrying capacity of the natural resources within their world, and years of experience.[37] It is from these sources that advice was sought for the development of *kapu* (prohibitions; sacredness) to maintain healthy fish stocks, forests, and farms.[38]

The chiefs and the *konohiki* lived through the labor of the people. Kamakau (1964, 8) reminds us, "Of the *maka'ainana* it was said that, in the end, the well-being (*pono*) of the kingdom was in their hands." Hundreds of wise sayings and stories pieced together from native tenants indicate the right time and seasons for gathering and leaving resources alone, where to locate plantings or not, how to keep fish stocks plentiful and close at hand, and the social behaviors that kept the system working (Pukui 1983). The taro farmer in the system saw his world and the actions he took as connected to and in relationship with all parts of the whole. He relied on conditions from the top of the mountain all the way down for the life-giving waters that fed his crop.

Management of an entire *ahupua'a*—from the mountains to the sea[39]—is a complex, full-time practice that has no real examples in modern-day Hawai'i. We have small-scale ecosystem and *lo'i* rehabilitation successes, and larger upland watershed protection areas, but are far from realizing recovery of the land divisions that embrace them. While some would argue otherwise, we simply no longer have the elements in place that made the model work. A fence in the watershed that fails to address increased ungulate pressure outside its boundaries in conjunction with invasive species regeneration inside, and fails to empower community comanagement, misses the basic principles of a functioning *ahupua'a*. Separating one component from another (imbalance) or selecting one practice as the primary solution distorts the outcome. For example, while the County of Maui has a "show me the water" ordinance for development meant to reduce pressure on already overtaxed aquifers, it does not yet have a matching riparian habitat set-aside plan to reconnect *mauka*-to-*makai* (upland to lowland) water corridors. It recommends water efficiency standards for new homes, but federal and state agencies lack effective counterpart standards and incentives for agricultural lands that would increase soil capture and moisture retention in the lowlands. Integrated, long-term groundwater and surface water recharge planning has yet to become a requisite part of sustainability. A healthy *ahupua'a* requires that watercourses be reconnected, including rehabilitating ancient *lo'i* and restyling urban areas to protect gulches and streams. It also obliges us to reforest midelevation corridors and recreate the "walking places" and aquifer capture sites for rain.

The strongest vestiges of whole systems management remain in Hawai'i's rural taro farming communities, even as the agencies that now hold jurisdiction over larger *ahupua'a* lands and waters have isolated themselves from this resource. An opportunity awaits state and county agencies to remove conflicting agendas, redundant staffing, programs, and expenses in reorganizing around *ahupua'a*. Like a microcosm of the world between the *lani* and the *kai uli*, taro farmers and the *'āina* they care for are caught out of balance in a traditional management system that depends on all our actions for well-being.

Mole: *Agriculture fit to the land*

From the previous *mole* of *Relationship,* it follows that the land (soils, water, vegetation) and the landscape (broader topography) dictate what type of agriculture and what food crops belong in a place, rather than the conventional notion that with the aid of a tractor and a bag of chemical inputs we can plow anything into production.

In 2004, E kūpaku ka 'āina, a small local nonprofit organization, framed its mission around restoring degraded natural habitats and "agriculture fit to the land." It defined that as a return to the roots of agriculture, a word that meant "a nurturing and fostering of the land." It was also an agriculture where land shaped the work and man belonged to the land, where practices and crops fit the local topography and resources of a place and the cycles of the seasons. Charlie and Paul Reppun, long-time taro farmers on O'ahu, have called this "endemic agriculture" to identify agriculture that evolves with a place. This includes sourcing soil inputs on-farm and in the surrounding *ahupua'a* (such as *limu* [seaweed], fish and bone meal, and mulch material) as much as possible and using "appropriate scale science and technology" (Oahu Resource Conservation and Development Council 2012; C. Reppun and P. Reppun, personal communication, 2012). Echoing the discussion on the moving target called "balance," the search for sustainability on-farm is a constant assessment of practices, crop choices, input sources, and ecosystem responses.

Hiʻohiʻona: Microtopography and microclimates matter

It is commonly understood in ecology that borders between ecosystems (edge effects) support a high rate of species biodiversity (Fagan, Cantrell, and Cosner 1999; Leopold 1933). In Hawaiʻi this played a role in the prolific endemism found in the islands. So, too, does a natural terrain provide niches and micro-habitats for a more diverse agriculture. Indigenous farming throughout the world is replete with examples of planting designs and practices developed specifically for the range of places people inhabit.

The Pacific islands of Futuna and Mangaia developed dry terraces and wet-land ponds for *kalo* production that look remarkably similar to those of Hawaiʻi. Where they diverge is in size. The two islands are both much smaller than any of the main Hawaiian Islands, and hence their field systems reflect local dimensions (Kirch 1994). Across the Hawaiian Islands, the largest taro fields are found on the broad, alluvial flats and wetlands at river mouths near the coast, such as Hanalei, Kauaʻi, or the front portion of Waipiʻo Valley, Hawaiʻi. Their banks are formed by mud, and they fit these wider spaces. Further inland, *loʻi* in Waipiʻo are smaller, with stone *kuāuna* (banks) formed from rock sourced near the river-bed. As elevations increase, valleys narrow and streams curve inland, *loʻi kalo* terraces hug the bends, sometimes becoming integral parts of the stream; they drop into depressions, rise more steeply from one to the next, shrink in size and lose their rectangular formality, shifting to parabolas. Huge boulders are integrated into the walls to withstand more concentrated water flows. An in-ward curve or an outward curve of a stream dictates the design of the *ʻauwai* that feed the patches as well as how water returns to the stream of origin or the intended receiving body (i.e., the ocean). The planner Ian McHarg (1969) called this "designing with nature,"[40] something ancient Hawaiians did with great skill.

At a second level, microtopography and microclimates influenced practices attentive to the natural character, and movement of soil, light, water, wind, and temperature (see "Practices matter," below). Hence, *kalo* in general might re-spond better under one planting method than another dependent on location. Additionally, planters of old selected specific *kalo*, *ʻuala* (sweet potato), *maiʻa* (banana), *kō* (sugarcane), and *ʻawa* (kava) cultivars most suitable to the microclimate of a site. They might choose the hardy, slow-growing Piko ʻeleʻele taro for the cold waters of rainy, upper Haipuaʻena or a Haehae (Piko uliuli) to withstand the fierce winds and warm lowland waters of hot Kāʻanapali or Waiʻanae; a Kāī uliuli taro for a deep mud patch, the fast-growing Lehua ʻeleʻele or 'Apuwai for the valleys with short light days; the long-seasoned 'Ohe for the cooler, rainfed fields of Kula; the Mana 'ele'ele for the terraces of dry Kahikinui.

Different topographies, soil types, slopes, sun exposures, elevations, and climates were all factors in choosing varieties and terrace system design. The reverse notion of making a site fit the crop or the cultivar requires more energy, more water, and holds greater risk for crop survival.

Hiʻohiʻona: Scale—size matters

Scale in agriculture is the one topic that tends to be avoided in Hawaiʻi because of our plantation history. Owning or having rights over large landholdings is common to both *aliʻi* and plantation eras. The plantations developed large fields mirrored after trends on the mainland.[41] To support this configuration water had to leave the streams, *ahupuaʻa*, and even the *moku* (district consisting of several *ahupuaʻa*) from which it originated. The result, as described in chapter 3 of this volume, is constant aquifer and soil erosion. Did Hawaiians have large-scale agriculture? Yes and no. In the ancient political system, *aliʻi* held sway over large districts, but the field systems within each *ahupuaʻa* tell a different story.

Large cultivation systems are found on open slopes, across broad stretches of coastal lowlands, and along the reaches of streams in well-shaped valleys throughout the islands. While it is true today that the largest individual fields are found in these same areas, it is not so clear that this was the case in ancient times (see chapter 3 of this volume for the impacts of immigrants on *loʻi* field design). Ellis (1825) describes stretches of *māla* (which he called plantations) and taro fields of several acres in size; however, on closer inspection, they are primarily in dryland regions. The lush growth he recorded in some districts masked the presence of numerous low walls, such as those in Leeward Kohala. The impressive agriculture system at this site covered more than fifty-five kilometers (34 miles) over thirty-three *ahupuaʻa* and 6,070 hectares (15,000 acres), but individual planting fields ranged from 0.28 to 1.37 hectares (0.69–3.4 acres) (Kagawa and Vitousek 2012; Ladefoged and Graves 2000; A. Kagawa, personal communication, 2011–2013). Kalokuokamaile, who lived in the adjacent district of Kona, describes eighteen field types of specific sizes, shapes, and locations (in Handy 1940, 49–51). The wetland patches of Kamehameha at ʻIole and the terrace called Umiloʻi, belonging to the chief Umi-a-Liloa, to the west of Waipiʻo bay, Hawaiʻi, have been described as "large," but their true measure is unknown (Handy 1940).

In Maui, ancient *loʻi* sites also indicate that while individual systems were large, the fields themselves were frequently small in size (one-eighth acre or less) and fit within the riparian landscape.[42] This makes sense in the narrow valleys, yet even in broad lowland districts such as Keʻanae-Wailuanui and Na Wai Ehā ("the four waters") the patches remained relatively small. Conservative reckoning for the *ahupuaʻa* of Waikapū estimated 1,400 *loʻi* over an expanse of 1,000 acres (Pellegrino qtd. in Lukens 2010). A 1995 study documented 339 *loʻi* on the alluvial flat of Wailunui, East Maui (Parsons 2008). The *loʻi* of Honokōhau were said to have numbered over 4,000 within the walls of this deep valley (S. N. Hoshino and W. Wood, personal communication, 2012).[43]

Where larger patches are described in historic record (i.e., Hanalei and the Mānā wetlands of Kauaʻi, the ʻEwa and Kona plains of Oʻahu), it appears associated with the transition period of the mid-1800s. An exception is noted in the written literature: "When the patch or pond is intended as a fish pond, the beds are made larger and bound together by various contrivances to prevent their flowing away as the water is necessarily much deeper to allow room for the fish and weeds on which they were fattened" (Queen Emma, n.d.). Today, commercial taro farm field sizes have not changed much, averaging one-eighth to one acre in size.

Logistically, caring for a big patch was easy if the labor was available; however, on a practical level, if the patch was too large, a farmer ran the risk of not being able to maintain his fields.[44] Water might also become too warm in a large patch, as noted by Handy (1940, 37) and increase the chance for disease and uneven production. Smaller fields are easier to manage, particularly under severe water constraints. Larger fields are tractor efficient. They can also handle distribution of alluvial soil deposits during floods better than small fields but are harder to relevel when that happens because of the sheer volume of mud that accumulates. The size of a field in the traditional system matched water and labor resources, as well as the landscape.

The conventional attitude that if one acre is good, a single one-hundred-acre farm is better just might not serve the goal of local food self-sufficiency and agricultural sustainability. Larger fields have led to machines doing the work of humans. It seems also to have encouraged large-scale, single-cultivar plantings, because that is also easier than managing dozens of crops or cultivars on a single farm. On the surface it appears that one person can do more and grow more; but it is also more costly for farmers (expensive machinery, fuel, chemical inputs and pest controls, maintenance, loans, and insurance) and the ecosystem (reduced capacity to retain organic matter and moisture, loss of niches, and diversity, soil erosion, and crop residue runoff). The financial and ecological debt is difficult to reverse. One hundred one-acre farms gets us closer to the goal.

Agriculture statistics in Hawai'i indicate a reproliferation of small farms in the last decade and that the growing diversity of food crops in local markets is found primarily on farms of ten acres or less (National Agriculture Statistics Service 2002–2013). This, along with local anecdotal evidence, suggests that small family farms and subsistence farms of a few acres or less reflect more diverse cropping systems and create far more jobs.[45] One argument for big farms is that the prices of inputs go down when purchased in large quantities. Cooperative purchasing of inputs can achieve the same result. A challenge with big farms is that too much of a single crop is ripe at the same time, and local markets cannot absorb the produce. The solution has been to ship produce to mainland and other markets, adding fuel-mile costs to the food. One farmer interviewed for this chapter articulated the dilemma: "A lot of things have conspired against the American farmer that forces them into a market driven model. Taro is successful because there is so little of it, and it is [consumed locally]." If one goal of sustainable agriculture is to feed ourselves first, we do not need to size ourselves out of our own markets.

Another aspect of size is in the produce itself. Size can be the result of breeding, nutrients, or planting methods. There are enough stories and photographs among *kūpuna* and archival material to suggest that the average size of *kalo* corms was much greater one hundred years ago than today.[46] There is also evidence that makers of *pa'i 'ai* favored the smaller *'ohā* (secondary corms) or the corms grown in the cold, upland waters for their higher concentration of starch and reduced sugars, while the mills sought the larger corms, but there is no way of knowing whether this was the norm (Pukui and M. D., n.d., historic photographs).[47]

Judge Hatch, speaking at a Mohonk Conference more than one hundred years ago, said, "The feeling in Hawaii is almost universal that the foundation of her prosperity is the small farm" (Board of Commissioners of Agriculture and Forestry 1904, 61). That is as true now as it was in Hatch's time.

Hi'ohi'ona: Practices matter

It follows that if agricultural systems were "fit to the land," that an array of practices would be employed to match the behavior of light, winds and rains, water, and the conditions of the soil at each site. This is evident under both wet- and dryland cultivation methods (Handy 1940; Handy, Handy, and Pukui

1991; Kalokuokamaile 1922; Iokepa, n.d.). Queen Emma (n.d.) wrote, "The culture [of *kalo*] is varied in different soils, different situations, localities and is regulated by a supply of water from aqueducts, springs, marshy lands, or by rain and dew only." The *ihu kū, pūpū kolea,* and *puʻepuʻe* (mounded planting methods) were adapted to both wetland cultivation and dry sites. These styles addressed low water flow and deeper, marshy conditions, but they also represent a progression (annual changes in planting practice) based on declining soil health after repeated plantings. The next step after *puʻepuʻe* was to fallow the land (Queen Emma, n.d.).

Soil renewal practices were also influenced by the resources that elevation and ecosystem provided. Kepelino and Kalokuokamaile recognized at least six planting methods specifically for dryland conditions, not including mulching practices (in Handy 1940, 47–53). The practice of *kuaiwi* (lines of stonework) was both a structural form and a mulching method where rocks dominated the landscape (Kelly 1983), such as the long-studied dry planting systems of Kohala. At this site, stone and earthen walls that followed the contour of the land held soil, caught moisture, and provided windbreaks, while those that ran *mauka* to *makai* (uphill to downhill) assisted in leading water-laden mists down the system.[48] The *māla ʻakōlea* and *ʻamaʻu* (fern gardens), the *pā kukui* (candlenut field; *Aleurites moluccana*), *pāhala* (pandanus field), the *ʻekī* (ti leaf patch; *Cordyline terminalis*), and *kīpulu* (mulched field) were both types of fields and methods of mulching using locally available materials. *ʻAmaʻumaʻu* (the young fronds of *Sadleria* spp.) fern was recorded as a preferred mulch throughout the islands. Kohala farmers relied heavily on *kō* (sugarcane) as both a windbreak and a moisture trap in fields where the famous ʻĀpaʻapaʻa wind tears across the slopes. It also provided a fast-growing and prolific source of dry material that was laid down as organic matter for the soil. In wetland sites, the *hau* (*Hibiscus tiliaceus*), *kukui,* and other plants played similar roles as a rapidly renewable mulch source. Queen Emma (n.d.) noted that "entirely volcanic [soils] require oxidization by exposure to air and sun before it is fertile."[49]

To reinvigorate the plants themselves, at least two practices are noted. Growing *kalo* in the mountains or at the edge of the forest strengthened taro stock.[50] Today, the rotation of *kalo* from *loʻi* to dry beds fulfills a similar role. Farmers of old marked *huli* (taro tops) to indicate unhealthy planting material (Kamakau 1976, 34) or threw the *makua* (parent stalk) away, planting only next generation *ʻohā* (shoots). The oldest generation of today's taro farmers tracked the number of times a parent stem was replanted using a similar method of marking or nicking the bottom of the *huli*. After the third planting, these *huli* were thrown away, ensuring a rotation of younger, healthier stock in the fields.

In all of this, one can see a keen attention to cyclic and long-term rebuilding of the soil and seed stock. Recent trends in local agriculture suggest a shift back toward soil husbandry; however, while indigenous microorganisms (IMO) and biochar in and of themselves are excellent inputs for reinvigorating soil microorganisms and processes, they require a balance of organic matter to make more mud in the *loʻi* or soil in the *māla,* a necessity where this has been neglected for so long.[51]

It is worthwhile to briefly juxtapose this place-based approach to agricultural productivity against the state's current assessment of prime agriculture lands. The Hawaiʻi Department of Agriculture's Important Agricultural Lands (IAL) rankings are derived from soil type, soil depth, and water availability data for a single commercial method of planting and two crops (sugar and pineapple). Remaining lands are designated for ranching, or game management, based on perceived low productivity when in reality it may be only a function of the wrong crop in the wrong place using the wrong practices.[52] The examples of dryland terracing in South Maui and North Hawaiʻi are spectacular examples of contrast to the IAL. That these lands are used only for ranching now is the result of our own misunderstandings and missing pieces in the system, all of which we are capable of restoring.

Internal practice also matters. Taro farmers had to be cautious, patient, aware, and attentive to the larger atmosphere, the ecosystem in which they lived, the immediate conditions of the farm, and the plants in their care. Knowledge of seasons, cycles, and ecosystem give-and-take all fell under farmer *kuleana* (responsibility). Whether a plant might be too aggressive (invasiveness) or need special tending, whether a pest or disease arrived when a number of conditions converged, or if there was too much pressure on a resource (e.g., fish or forests) had to be observed and mitigated. Relating back to the concept of relationship, a stress in the environment needed a paired response of giving back (healing).[53]

Because the Hawaiian planter managed his farm both for the immediate and long-term future, he was also proactive. This allowed for the production of *kalo* and other crops in the same locale over and over again. Taro farmer Kyle Nakanelua describes true "Hawaiian time" as purposeful. You planned far in advance, sometimes years ahead, to maintain farm fertility, to provide for a big event, and to feed future family members.[54] In ancient Hawaiʻi that timeline was sometimes centuries long. While these "best management practices" did not always happen, particularly as society changed, they were enough in use to have maintained food self-sufficiency in the islands through World War II.

Iluna no ka ua, waele 'ē ka pulu. Remove the mulch while the rain is in the sky. Prepare for rain before it comes.[55]

Mole: *Resilience*

It holds that if a farmer looks to the characteristics of the landscape in the place he (or she) lives to shape the type of agriculture he implements, he would also develop practices with a high degree of adaptability and flexibility to weather natural events, a form of long-term stability. This seems counter to current definitions of "stability" (firmness in position, resistance to change) until one observes the details behind the lasting presence of taro farming. Two trends are apparent.

The first is designs that did not oppose natural processes and elements. For example, to handle big water in a stream, a farmer builds a breakaway rock dam to carry water to his *lo'i*. If it comes apart during a storm, he can rebuild it with minimal effort. The design also allows native fish and *hīhīwai* (*Neritina granosa*) to traverse the dam with ease. A farmer of old might have used *māmane* (*Sophora chrysophylla*) or *'ohai papa* (*Sesbania tomentosa*) branches to settle the mud after a storm. The combination of leaves and silt would have yielded a nitrogen and amino acid–rich mulch.[56] To build strong plants, you were admonished to "pluck the leaves two to three times to force strong rooting" (Queen Emma, n.d.).[57] Moving fields when the river moves and changing practices and varieties when soil or weather conditions shift are part of the toolbag of a resilient agriculture.

The second trend is maintaining a high degree of diversity and applying that diversity in a variety of ways. The financial saying "don't put all your eggs in one basket" originated on the farm.

Hi'ohi'ona: Crop and cultivar diversity

Farmers typically reduce risk by creating diversity across locations and crops, and within crop cultivars (Tadesse and Blank 2003). These three aspects of diversity describe the Hawaiian farmers' strategy perfectly. Monocropping is the antithesis of resilient farming, opting instead for planting exact replicas of a single cultivar across all locations.

A food supply or seed bank can survive storms, tsunami, sea-level rise, volcanic activity, drought, pests, or disease where a range of planting locations that differ in climate, water source, topography, elevation, aspect, and soil type exist, even where such threats are widespread.[58]

Crop diversity, or a multitude of crops, plant heights, and forms within a single farm, mimics a robust ecosystem. Depending on design, it can have structural strength above and below ground that enables it to withstand storms and high winds. A diversity of plants hosts beneficial insects, inhibits the spread of pests and disease, creates opportunities for integrated farm management and locally sourced inputs, and produces an array of food and use products. If one crop fails, there are others to rely on.

At the apex of agricultural crop diversity are examples such as the traditional forest gardens of Southeast Asia, which draw on surrounding rainforest biodiversity and intentionally support hundreds of useful plant species within a single small farm unit (Levin and Panyakul 1993; Levin 1990). Species diversity was somewhat more limited in traditional Hawaiian gardens by comparison; however, they were described by early explorers and multiple visitors as having numerous plants integrated on the banks and borders of the taro or *'uala* (sweet potato) fields, including each of the staple foods described by Handy, Handy, and Pukui in *Native Planters in Old Hawaii* (1991). At least one taro farm on Oʻahu currently supports as many as seventy species of edible plants in addition to numerous *kalo* varieties, not including plants for lumber, cordage, medicine, and organic matter (i.e., weeds). Rotations of crops, animals, and fallow in each area of the farm allows soils to rest and replenish. Handy (1940, 143) illustrated how the limits of location and time could be expanded by growing more than one crop, writing that "the sweet potato can be grown in much less favorable localities [than taro], both with respect to sun and soil; it matures in 3 to 6 months (as against 9 to 18 months for taro); and it requires much less labor in planting and care in cultivation."

Finally, cultivar diversity is varietal richness within a single species. This begins with crop wild relatives (wild ancestors of domesticated plants) and is expanded by the formation of land races (local varieties developed largely by natural adaptation to the environment in which they live, as well as somatic mutation) and varieties formed by deliberate selective breeding.[59] The presence of many varieties within a field is an old strategy for reducing the risk of crop failure. It also makes for more robust genetic options for species survival.

What the original Hawaiian crop complex may have lacked in crop diversity, it made up for in cultivar diversity. Historically, the repertoire of indigenous food cultivars in Hawai'i included an estimated 300–400 varieties of *kalo*, possibly 150–250 varieties of *'uala*, forty varieties of *mai'a*, forty to sixty strains of *kō*, thirty-five or more *'awa* cultivars, perhaps five varieties of yam and the same of coconut, and a single type of *'ulu* (breadfruit).[60] This diversity allowed the traditional farmer the flexibility to grow food in a multitude of places. A farmer would carry with him to a new planting site "a sufficient number of [*kalo*] shoots of the varieties known from experience to be best adapted to the location provided" (Queen Emma, n.d.). Old photographs of *kalo* fields from the late 1800s through the early 1900s illuminate our understanding that *kalo* varieties were planted both separately and mixed together. As recently as the 1960s at least a dozen *kalo* varieties were still well known in the market. The dominance of Maui Lehua as the poi taro of choice since the 1980s is a loss for consumers and farm resilience.

Scientists argue that, despite the numerous taro cultivars developed in Hawai'i, the gene pool is rather narrow and hence places Hawaiian taro varieties at greater risk than their counterparts in Southeast Asia (Rao et al. 2010). Genetic weakness cannot be assumed in Hawaiian taros without a fair accounting of the loss of traditional soil care practices to taro health over the last 150 years. In recent back-crossing experiments summarized by Miyasaka (2013), it was found that Hawaiian taros carry the genes for disease resistance assumed to be absent by earlier research.

Currently, perhaps fifty-five to sixty Hawaiian taro cultivars have survived, nineteen *mai'a*, thirteen varieties of *'awa*, and thirty to thirty-five varieties of *kō*, along with an unknown number of *'uala* (A. Kagawa, personal communication, 2011–2013; Kepler and Rust 2011; Johnston and Rogers 2006).[61] This is still a substantial number with which to improve on-farm diversity and provide us with crop and climatic resilience and pest and disease resistance. With such richness at our fingertips, one farmer asks, "Why are we trying to become an Iowa corn state when we have something so unique?"

Hi'ohi'ona: Multipurposed

One of the tenets of permaculture, an agricultural system informed by indigenous practice, is that everything on the farm should have multiple purposes for greater efficiency, from roads and structures to animals and plants (Mollison 1988). The dual function of taro systems as connectors to streams and habitat for waterbirds is an obvious example. *Kalo* is not just a food and an embodiment of the gods. It is substitute for a number of fish in ceremonial offerings, a dye source, and medicine, as well as food for animals and fish at the *koʻa* (coral head; fishing ground) in the ocean. Historically, it was used to pay taxes to the chiefs and is just as important today in social reciprocity and trade. As a food, *kalo* also presents a wide variety of options for value-added products from the traditional *kūlolo* (a steamed dessert) to taro chips, restaurant dishes, and *paʻi ʻai* pizza dough, diversifying the end use of the *kalo* and the jobs that gather around it. *Kalo* was and still is a teacher and an ecosystem indicator for the well-being of the *ahupuaʻa* and the community.

Mole: *Locally sustainable and renewable*

Natural resources that have replenishment cycles counted in centuries, or that are unlikely to be created ever again, are finite assets in our lifetime. Old-growth *ʻōhiʻa* trees (*Metrosideros polymorpha*), springs and aquifers, cinder, stones, coral, soil, and sand fall into this category. Traditional culture tells us that we should use their services, and wisely, rather than use up their physical form.

Maui has overdrawn its water accounts in ʻIao and Waikapū districts.[62] Our current and projected rate of water use, coupled with the latest drought cycles, places freshwater aquifers in the category of a nonrenewable resource. We have talked about water cycles in the discussion of interconnected relationships, so why continue the discussion here? Kapua Sproat (2010, 187–194) describes the historical scope of what has happened to stream flows in Hawaiʻi, but she chooses to begin with a stanza from the chant *He mele no Kāne* (A song for [the god] Kāne) called *Aia i hea ka wai a Kāne?* (Where are the waters of Kāne?). The chant speaks about the abundant water cycle and where to find water in the islands—in the dew, the mists, the rain, springs, and streams. If we learn from traditional observation and shift the way we understand this resource, as well as how to recreate places for it to gather again, water becomes not in our generation but perhaps in the next, a resilient and renewable resource.

If water is our most necessary resource for life, then soil is the sponge that holds it and the medium that allows us to grow food. We have spent the last two centuries turning our heads away from the loss of topsoil into our oceans around Hawaiʻi. The dust clouds and soil runoff at Kīhei and Lāhainā are but two examples that represent an unrecoverable loss to the system, like a bank account that has been overdrawn.[63] The interest on the debt will take centuries to pay back. But again, if we understand the mechanisms for rebuilding and retaining soil, we change over time from a finite to a renewable resource that increases moisture retention capacity and the regenerative breath of the land itself.

Hiʻohiʻona: Locally sourced inputs, revisioning resources

Importing fertilizer and fuel is a losing game for farmers and agricultural sustainability. The mining of fossil fuels, minerals, green sand, coral, gypsum, or other elements in soil amendments, organic or not, from another state or country is not sustainable—on a local or global level. The costs to transport these amendments subtract further from the sustainability equation.

To reverse this *requires* that we shift to locally sourced inputs. The outcome is likely to bring significant change to soil practices for both conventional and organic farmers. But do we wait until we have no choice (i.e., when the cost to import amendments and fertilizers is too high, or the source runs out)? Or do we become proactive and relearn these traditions now while we can still bridge current practices and have some buffer for trial and error? The definition of sustainable agriculture is also being shaped by consumers. The desire for chemical-free food choices is added incentive to make the shift.

Traditional farming provides us with a well-populated list of plant sources that were used to build soil productivity and health, including *hau, kukui, hala* (pandanus), *ʻulu,* reeds, and grasses, banana, *awapuhi* (shampoo ginger), ti leaf, ferns, and much more.[64] Some of these plants are considered invasive species now only because they are no longer tended as mulch crops. The *hau* and *kukui,* in particular, are abundant producers of organic material with multiple uses. While only minimally in use today, Hawaiians deliberately planted and coppiced these species in a sustainable manner to feed the soil. These two species may also have been applied before or after periods of blight to prevent or reduce establishment of fungal pathogens.[65]

Alien plant species provide a tremendous source of biomass as well. Without control efforts, alien species will continue to advance into native habitats. Using this resource involves some learning about invasiveness and how to control species so that biomass sources can be managed by local communities in a manner that prevents regeneration. It is an opportunity for local *ahupuaʻa* restoration and agricultural self-sufficiency.[66] A good example of recycling invasive species is in the application of alien *limu* (seaweed) that washes up on shore or is removed from coral reefs in Hawaiʻi. *Limu* is a source of calcium, micronutrients, trace minerals, and proteins. It contains alginic acid, a soil conditioner that improves structure and moisture retention capacity (Stephenson

1968). To date, we have no record of whether Hawaiians used this resource in their plantings. We do know from early accounts they burned their dryland fields with a covering of grasses, breadfruit, or *hala* leaves and that the lime from the fires gave the young *kalo* a burst of growth that allowed it to achieve heights of seven to eight feet.

Recycling plays a major role in developing local inputs. Many old-time taro farmers dried the peelings of cooked taro and added them back into the soil. Fish guts, animal bones, shellfish, eggshells, food leftovers, tree trimmings, moldy hay, and composted manures are just a few of the hundreds of resources at our fingertips.[67]

Chemical fertilizers rob the future in a second way: they make no new soil; only organic matter can do that. Part of the strategy in developing local soil amendment resources includes devoting enough land in agriculture to long-term fallow—anywhere from one to thirty years, depending on soil conditions. Instead of "abandoned" fallow, the opportunity to grow and coppice select trees or other plants as a managed, renewable supply of organic matter is also an opportunity to increase infiltration and reduce soil loss via runoff. The ratio of fallow "supporting lands" to actively planted lands is something to plan into the future.

Locally sourced energy and fuel is another piece of the puzzle. In 2009, 43 farms used wind energy and 520 used solar panels to meet their energy needs; this represented an average savings of $2,215 per farm on utility bills (Hudson 2012).[68] Subsidies and programs to assist farmers in the shift towards alternative energy will not last much longer but are worth tapping into while they still exist.

Hiʻohiʻona: Food here first

Hawaiʻi imports 85–90 percent of its food. We ship our fish, vegetables, and fruits to California and Japan before we feed ourselves. It seems absurd that we import the very same things we already grow, or can grow, locally. How do we reach abundance if the practice of sustainable agriculture remains connected to a model that constantly pushes the system to produce more—and more for export—when in fact our islands have limits? We cannot gain an accurate measure of abundance if we are not feeding ourselves before we measure surplus.

Sustainable agriculture also encompasses the consumer side of the equation. Educating residents, wholesalers, grocery stores, chefs, restaurant staff, and visitors about the value of sourcing locally and the beauty of crop seasonality steers us away from the demand that all foods be available all the time—a way of thinking about food that increases our dependency on imports. Anticipation of the seasonal arrival of fruits and vegetables allows us to savor and appreciate our changing diets and gives us reason to celebrate. Already, communities have found this creates opportunities for local events and value-added tasting experiences that generate new income.[69]

Looking at the big picture of food production, Reppun in *The Value of Hawai'i* (qtd. in Howes and Osorio 2010, 39–46) concludes that 140,000 acres would have to be committed to fruit and vegetable production to meet local annual consumption, including approximately 30,000 acres for taro (more than fifty times the existing lands in commercial *kalo* production).[70] To meet the protein needs of Hawai'i's 1.3 million people using current strategies for beef production would require "a whopping 6 million acres (or 1.9 million more than the archipelago has)." That suggests that the restoration of Hawaiian fishponds will be necessary as an alternate protein source.[71] The vision of the students who participated in the data collection and forecasts for the study was to delink from our islands' dependence on imported factory-farmed food and reduce our carbon footprint, a significant portion of which comes from shipping in food and agriculture inputs to the islands. They also saw an opportunity to increase local jobs in agriculture for the unemployed.

Hi'ohi'ona: Employs many and employs locally

Large-scale agriculture created an import of another kind in the islands. The makeup of Hawai'i's residents is a direct result of imported labor to the plantations over the last 150 years. Faced with rising costs of living and the lack of quality jobs, young, talented local residents have been leaving the islands in a decades-long exodus. Sustainable agriculture has the potential to create thousands of jobs in a diversity of sectors, from on-the-ground food production to research; add value to existing careers; and support resources and education. The development of locally sourced, renewable soil amendments is a wide-open sector for new entrepreneurs and jobs. Hawaiian food cultivar diversity is an opportunity for each valley and farm to rediscover and specialize in the varieties that once grew there. In cultivating the palates of consumers, so too will we cultivate new jobs.

Several factors influence a return to the land in the taro-farming community: a strong desire to come home to the places, practices, or things that people grew up with; a desire to reconnect to culture; the availability of land; the ability to live on the farm, which reduces overall family expenses; and a fair price at market for their goods.

A comparison between large- and small-scale (direct) farm jobs in Hawai'i reveals the potential for increased employment as we move in the direction of sustainable agriculture in the islands. Using the Ladefoged et al. model (2009), traditional agriculture created roughly one job per acre and provided food self-sufficiency. This is a ratio that remains relatively constant on the smallest farms.[72] The sugar industry employed roughly 12.5 people per one hundred acres (Decision Analysts Hawai'i, Inc. 2009). For their diversity of food and use plants (including herbs and spices, medicines, cordage, building materials, and much more), planting and management methods, and value-added and niche markets, small farms represent the best kind of farming. They require more human labor and are less dependent on fossil fuels and machinery. They also depend on a strong relationship between the farmer, the food grown, and the communities they feed. One secondary outcome of that set of connections is that money tends to stay and circulate longer within our communities before it leaves the islands (Schwartz 2009).

Mole: *Leads to abundance*

Always we should ask, "What are we sustaining?" and "Where does that lead us?" Are we interested in the status quo, or can we move into a position of abundance and, if so, how do we define that?

"Abundance" is a copious supply, a profusion; a degree of affluence (wealth), and also a fullness or benevolence. The word appeared in its English form sometime in the mid-fourteenth century derived from the Latin *abundantia* (fullness) and *abundare* (abound, to overflow). In Hawaiian, the word *waiwai* (wealth) is defined in the dictionary primarily as physical riches. Pukui suggests this use of *wai* "is related to [the meaning], to retain, rather than to . . . water [*wai*]" (Pukui and Elbert 1986, 380). Yet, it cannot go unnoticed that in ancient Hawai'i the acquisition of resources, whether by inheritance, association, war, or labor, was associated with an abundance of food, which could not be achieved without plentiful *wai*, or water. In interviews in 1975, Pukui affirmed this deeper understanding and expands on it, describing the relationship between *punawai* (spring) and *kūpuna* (elders) as a source of knowledge (wealth) for the next generations (Maly 2003). As discussed earlier, an accumulation or surplus of resources did not necessarily beget goodwill in ancient Hawai'i and was often met with some form of reckoning from the gods. Generosity, the second part of the English language definition, implies a different way of looking at the world.

Andrews ([1865] 2003) refers us to the words *he lako, he nui wale* to express the concept of abundance and defines this as possession of what is necessary for any purpose; a supply; a fullness; a sufficiency; to be great; to increase in size or magnitude. Pukui and Elbert (1986, 200) records the saying *he moku lē'ia*, a district of abundance, bringing us back to a connection with *'āina*. Far more poetic ways to express abundance are found within the stories, chants, and *mele* (songs) preserved in archives and Hawaiian language newspapers. *Kalo* farming, in the past and the present, is much more than a measure of income or jobs. In its traditional context it was a livelihood of abundance and well-being.

The most critical aspect of getting to abundance is one that is difficult for many to consider—population and the very real physical limits of the Hawaiian Islands. Assumptions of economic "growth" are predicated on "more"—more people, more purchasing power, more goods, more homes, and so on. Our unswerving belief in an economic definition of growth has created an amnesia about carrying capacity. What does that have to do with sustainable agriculture?

Large-scale agriculture that overtaxes the land and leaves it barren is not just privileged to the industrial era. It is also a function of population. Overburdened food production systems and environmental degradation together have a well-documented, probably causal link to the collapse of major civilizations (Diamond 2011) and landscapes (e.g., Mayan, Uruk/Sumeria, the US Midwest in the 1930s). Three trends tip the scale for land and natural resources: agricultural practices that ignore ecological principles and boundaries; a continuously growing population, whether by births or by in-migration; and a continuously growing outside demand for locally limited resources.

Key questions we should be asking within the framework of sustainable agriculture include not just how many businesses or jobs exist (or are projected) in agriculture but what kinds of agriculture exist, and whether they fit this place. What crops, where, and how? Not just alternative energy, but in what form and for whom? Where to build and where not; and the much harder issue of understanding our islands' limits and how to address them, so that our natural resources have a fair chance at recovery.[73] We need to make thoughtful, proactive choices to address these questions.

Hiʻohiʻona: Replenishing

If we draw from the foundations and elements that sustained *kalo* farming, we come away with an understanding that a sustainable agriculture is one that is constantly replenishing the resources that support it. It sets aside lands for fallow and for production of organic matter. It restores the surrounding ecosystem. It repairs or replaces, in place, what it has. And it fulfills us.

We have for years looked at natural resources and our local markets as something to divide up like pieces of a pie. As the plate has emptied, we grumble more and worry more. We put mechanisms in place to limit abuse and overuse, yet none of these mechanisms regenerate the pie. We are fighting over the crumbs rather than choosing proactively to make more pie. To reach sustainability means planning into daily practice and long-term action an element of putting back more than we take to replenish the resources we do have. Reconnecting streams, *kalo* systems, and fishponds will grow more food and more fish. Advancing native forests back down our slopes will grow more water. Growing mulch will improve our soils for more food, more jobs, and better health. Feeding ourselves first will reduce fossil fuel use and carbon footprints and the influence of volatile markets on the cost of our food and on food security. Diversifying the *kalo* varieties in our fields will create crop resilience and new consumers. Some things will be simple; others will take a long time. Like farmers, we need to cultivate patience.

Abundance, however, is immediately available to us. It is a matter of changing perspective and reconnecting to *ʻāina*. It gives us room to laugh more and enjoy life.

If agriculture is to be truly sustainable, it must not only replenish the ground it grows from but also replace aging farmers with new blood. The $0.67/lb farmgate price of taro is less than encouraging for young families that need to support themselves with a dignified income. The typical answer has been to diversify or value-add. A new renaissance has drawn on the traditions of old with a return to hand pounding *kalo*. *Paʻi ʻai* (hand-pounded poi) requires a higher-quality *kalo*, which has in turn raised the value of the crop at farm gate for this niche market and recreated a demand for the old cultivars as well. An increase in direct farm-to-consumer and farm-to-chef markets provides a second alternative. The growing social capital movement, including farm-based, locally sourced pop-up "restaurants," has fully embraced food fresh from the farm.[74] Each is restoring prestige to being a farmer and increasing the choices farmers have for making a living.

Hiʻohiʻona: Celebrates and maintains connection to place

Kalo is never just food, a vegetable, or commodity. Even if a farmer today is not moved by the culture, history, and knowledge embedded in this plant, the consumers on the other end may see it as their connection to who they are or as the source of their well-being. When a farmer grows *kalo,* that plant retains the genetics of the first corms that came ashore more than 1,000 years ago. Each time he bends to plant a *huli* in the mud, his movement echoes that of the very first planters in the islands. Hāloa lives through the perpetuation of this beloved livelihood. Taro-farming communities celebrate the self-sufficiency and resilience that *kalo* represents whenever they gather around the plant, whether through festivals or the making of poi. *Kalo* has come to represent the face of Hawaiʻi, and it retains a strong connection to culture.

If the planter reaches back into the Hawaiian traditions and place names that surround him, he learns about the land that he farms in ways that a soil report or a land survey could never do. The writer Dennis Hirada (2000) describes his sense of discovery in reconnecting to place:

> Like a riddle, each place name and wind name challenges the reader to figure out why it was given, and leads to discoveries about the special character of that which is named. Such knowledge of geography and weather was essential to the success of those who lived close to the land and sea and depended on them for their survival and well-being. I had traveled up and down this windward coast of Oʻahu many times in childhood; but only after I became familiar with the meanings of the place names and wind names did I begin to see the land in all its complexities, full of life and life-giving. Pahonu became more important and precious to me than Paris.

Hiʻohiʻona: Productive

Production in agriculture is typically measured in yield per acre. Despite the fact that *kalo* represents a very small portion of agricultural lands in the state (less than 1 percent), its production rates indicate it is one of the highest yielding food systems in Hawaiʻi. In 2008, it provided an estimated 12.9 million local plates with a healthy starch from less than 400 acres of commercial farms (Taro Security and Purity Task Force 2009).

The state of Hawaiʻi and the nation offer little recognition or support for what they call "subsistence" agriculture, mostly because on a purely revenue generating level, it does not appear to contribute to gross national product or county and state incomes. But there are other ways to measure productivity on the farm including product quality and diversity, soil health, self-sufficiency, personal health and well-being, and community relationships.

When we begin to look at monies not spent, productivity is represented by the savings to family, county, and state budgets. Health care costs go down when a family eats well and is more active. Subsidized housing and health care, unemployment, and Food Stamp costs go down when a family can feed itself and live on the land where it grows food. A family frees up cash to apply toward other things when it produces its own food. From this perspective, the value of subsistence agriculture to the state grows considerably. Small family farms in general provide a high degree of food self-sufficiency, and *kalo* farmers are no exception.

At some point, dollars not spent, ecological costs not incurred, and imports not needed (fuel miles and production costs) must be added to the definition and economic accounting of sustainable agriculture in Hawaiʻi.

Hiʻohiʻona: Growth and *kuleana*

We have grown up with an agriculture model shaped by the economic myth that perpetual growth exists, and this myth drives the market. A farmer only needs to look to the ecosystem around him to know that nothing in the real world functions that way: things grow and die, and something new is regenerated from the soil or the seed. He also knows that if he takes too much from the system, he will either have to put back or it will also diminish and eventually die.

Four decades ago, the economist E. F. Schumacher (1973, 34) proffered the idea that "small is beautiful." He suggested that a large economy could not be successful in the broader context of its demands on the environment. In a quote that mirrors the present, he wrote, "Wisdom demands a new orientation of science and technology towards the organic, the gentle, the non-violent, the elegant and beautiful." Essential to this model is the concept of "enoughness," its premise, to maximize well-being while minimizing consumption (needs and desires) (Bhongmakapat 2011). This runs counter to growth-driven agriculture. Schumacher would have pointed out that a healthy agriculture economy would differentiate between renewable and nonrenewable resources, a distinction that would make it more robust.

Kuleana (responsibility) is the final part of the equation for sustainable agriculture. In order for agriculture to be truly viable, it must also be supported by consumers and decision makers, and not just by buying and supporting locally grown produce.

In 2009, the Taro Security and Punty Task Force wrote,

> In the traditional system, it was the responsibility of the konohiki [headman] and the *ali'i* [the chiefs] to ensure that the maka'āinana (those who cared for the land) were cared for in turn. They depended on the farmer to provide for everyone who did not grow their own food. Taro farmers hold up their end of the *kuleana* by caring for the water, the land and the *kalo*, but what do they get in return? For today's consumer to expect cheap poi on the shelves all the time, for today's visitor industry to expect that the beautiful viewplanes of the taro lands of Hanalei, Ke'anae and Waipi'o will always be there, or lo'i will be available to visit, for the restaurant and raw food buyers to expect the farmer to produce a consistent quality product without providing active support for that to happen is out of sync with the reciprocity that is required of this lifestyle and for the taro to survive. (17)

When taro farmers have to come out of their *lo'i* to make right those things that they should not have to fight for (such as the return of water to their streams), as one farmer put it, "this is what is disrupting the balance.... Everyone has to take care of their side of this *kuleana* so that we can continue to care for the *kalo* and for those who eat our food."

RESTORING SUSTAINABLE AGRICULTURE IN HAWAI'I: RENEWING THE BREATH OF HĀLOA

There is no question that we have the land, resources, and knowledge to be food self-sufficient and restore sustainable agriculture in Hawai'i. We have already asked whether we wait until we have no choice and struggle through a decade or two of food poverty or if we will be proactive and help people get back on the land to farm now while we still have a buffer as we move away from dependent agriculture. The other side of that question is, "Do we have the will?" Are the owners and managers of our largest stretches of fallow agricultural lands (public and private) willing to allow change to happen and to be part of that change?

In 2010, the economic futurist Gerald Celente pointed out from his own perspective that all things are connected and that innovation, quality, and a return to small rural farms would be the way out of our current economic crisis. The following year, he urged us to develop clean foods and said that staying healthy in a poisoned world would be a priority. In 2012 he continued to say grow food; have a "revolution of the heart," and do what you love to do; have the freedom to think, be yourself, take risks, be innovative, have the courage, and the passion to return to abundance.

The *kūpuna* have done a great deal of work for us already. They have left clues in names on the landscape that tell us how the winds and rains behave, or how a locality or stream may be a flood zone or a place of bounty. They have left us stonework masterfully engineered to handle the behavior of water at each site. To understand this is to understand that years of observation, design experimentation, and experience occurred prior to the construction of the ancient agricultural systems that sit before us. All the tools are here.

As a group of islands 2,500 miles from the nearest landmass, it is critical that we listen deeply to this *'āina* that surrounds us for guidance. We have been gingerly stepping around the edges of our role in the procreative dance of growing food. Planting *kalo* is not about living in a bygone era. It is about bringing the knowledge of the past into an abundant future. In the valleys can be heard a *kani* (call) for a new generation to repair the ancient walls, turn the mud, and raise the poi pounder once again to *ku'i 'ai*.

E ala! E alu! E kuilima! Up! Together! Join hands!

NOTES

1. A gourd or wooden container, or calabash, was kept in most Hawaiian households to store poi, the staple food of the islands.

2. See *Hawaiian Dictionary* (Pukui and Elbert 1986, 340–341).

3. These cases have now made their way to the Supreme Court and are ongoing.

4. This increase is noted by 'Onipa'a Nā Hui Kalo, a statewide organization of taro growers and enthusiasts, and members of the legislatively established Taro Security and Purity Task Force. Hawai'i Department of Agriculture statistics do not reflect either the increases in total commercial and subsistence taro acreage or the number of farms. By one 2012 estimate from members of these organizations more than 106 active taro growers can be found on the island of Maui alone.

5. The School Gardens Program is successfully bringing gardening, food self-sufficiency, and fresh, healthy food to grades K–12 and beyond.

6. Taro varieties are being distributed from collections and farms around the state to hundreds of people each year. Led by Jerry Konanui, the number of and attendance at taro varieties identification workshops in the islands has multiplied fourfold in the last five years. The number of sites caretaking the Hawaiian kalo varieties has expanded from eight to fourteen in the same period. In 2013, close to 20,000 *huli* (planting stock) of more than forty cultivars were distributed to hundreds of growers statewide from just four collections (E kūpaku ka 'āina, He Pili Wehena 'Ole O Na Kalo program).

7. NASS statistics from 2002–2013 record a high of 5,800 tons processed taro (poi) in 2002 to a low of 3,800 tons in 2010, with fluctuations around an average of 4,400 tons. Major storm and flooding events have a typical two-year crop recovery cycle. The majority of fluctuations during this period can be attributed to such events.

8. From a representative sample of twenty *ku'i 'ai* practitioners who have also taught others to pound poi and to make stones and boards between 2007 and 2012—these are conservative estimates. Those numbers will have more than tripled by the time of this publication.

9. Farm-gate prices for raw taro average $0.67/lb at the mill (2013), the most common outlet for raw taro. At farmers' markets, growers receive $1–2/lb direct to consumers. The newest market for hand pounding provides growers with $2/lb farm-gate for high-quality taro. From the early 1800s to about 1910, all commercial poi was done by hand. Machine-milled poi first made the scene at the turn of the twentieth century and quickly replaced hand-pounded poi. The last handmade poi sold commercially was in the 1930s (Olszewski 2000). Health regulations initially restricted and eventually ended the practice, making it illegal to distribute to consumers. By the 1970s, while a few rural families continued the tradition of making their own poi with board and stone at home, very few others knew how; the majority had switched to grinders to make the poi. The cultural revival of that era inspired *kūpuna* (elders) from within the Hawaiian community, such as Uncle Eddie Kaanana and Uncle Harry Mitchell, to share the art of stone and board making and of pounding poi once again.

10. Discussions on the Hawaiian meaning of *pono* are relevant here.

11. Hawaiians recognized a complex taxonomy of stones. One type, *pōhaku lū'au,* notably named for the taro leaf (literally, cooked taro greens stone), was a dark, fine-grained waterworn basalt used in adze making. Considered by some to reference the dark color of cooked taro greens, it also referred to a soft porous stone used in the *imu* (ground oven) and hence for cooking taro and taro leaf (Pukui and Elbert 1986; Andrews [1865] 2003).

12. There are many versions of this story among northeastern Native American tribes and from family to family, some referencing only the corn or other food plants, including tobacco.

13. In Eliade's writing, "religious" was synonymous with "sacred" and should not be construed to mean related to a specific belief system such as Christianity, Judaism, or Buddhism. For some Hawaiians, cultural practice is also religious practice.

14. In Prosek (2010), Kelly Davis speaks from an indigenous perspective about science: "Western science is observation and recording . . . [and] indigenous science is oral. It's not one man's lifetime of knowledge. It's a continuum." Observation and recording (whether written or oral) are key to both. As Davis points out, it is the generations of contributors (longevity, replications) and complexity of observations that differentiate the two.

15. *Ao* is a complex word and set of concepts. The reader is directed to the *Hawaiian Dictionary* (Pukui and Elbert 1986) for further insight.

16. Both have myriad forms. Readers are directed to Beckwith's *Hawaiian Mythology* (1985). For more in-depth reading, see Beckwith (1951) and Johnson (1981) for an introduction to Kū and Hina, the pairs of organisms and their order of birth.

17. Examples are found among native plants at particular sites, such as coastal *'ilima papa*, a prostrate form of *Sida fallax*, the *naio papa* (*Myoporum sandwicense*), a dwarf, low-statured false sandalwood sometimes found on impoverished trachite soils, the stunted *'ōhi'a papa* (*Metrosideros polymorpha*) limited in stature by boggy soils, and the upland *maile kūhonua* (*maile* sprouting from the earth; *Alyxia* sp.), the name indicating a young, erect stage of this fragrant plant before it begins to climb and vine.

18. Literally and figuratively.

19. 'Ōlelo No'eau no. 531.

20. Modern-day efforts to restore water to Hawai'i's streams take place on the farm, in the community, in protests, in the media, and the courts and has been likened to a David and Goliath battle. Current contested cases have been ongoing since the late 1970s, yet petitions to the king and the courts protesting the taking of massive amounts of water out of streams and districts for sugar at the expense of taro farming began almost 150 years earlier. And still, taro farmers persevere.

21. There are always exceptions to any such statements; however, this theme is found repeatedly in Hawaiian literature in both English- and Hawaiian-language sources.

22. Stories such as the removal of *'o'opu* (goby fish) from Wailau stream, Moloka'i, by the god 'Ai'ai as a punishment to the people there for overfishing (Kawaharada 1992) and the jealous covering of Puna with lava by the goddess Pele after a chief boasted of the beauty and lushness of the district (Pukui and Curtis 1971) are but two examples.

23. *Pili* is the first stage in poi pounding with the taro beginning to stick (Pukui and Elbert 1986).

24. This pertains to water velocity and volume. Floodwaters and sediments can contain excess nitrogen loads that do impact taro corm quality, but also renew nutrients in fallow fields.

25. The loss of taro farms (and fishponds) in these zones has played an enormous but substantially unrecognized role in the decline of Hawai'i's wetland waterbird populations.

26. This planting method can be observed at Amy Greenwell Ethnobotanical Garden in Kailua-Kona, Hawai'i.

27. *Hahai no ka ua i ka ululā'au*, rain always follows the forest (Pukui 1983, 'Ōlelo No'eau no. 405). This proverb describes how the mists and rains "walk" upon the forest canopy, which allowed precipitation to descend to farmlands in the lower elevations. On Hawai'i island a midelevation *'ulu* (breadfruit) forest assisted this process; in East Maui, a band of *'ōhi'a 'ai* (mountain apple) played a similar role.

28. Taro Security and Purity Task Force 2010 report to the legislature and personal communication with taro farmers.

29. The recommended average temperature is no higher than 78°F (Evans 2008).

30. In addition to bacterial growth from low stream flows, other sources include over-fertilized fields (wet and dry), wash water, piggeries, stables, aquaculture facilities, and road

surfaces that drain into the same water sources used by taro farms. Excess nitrogen increases the frequency of soft rots and *loliloli* (a rubbery, watery condition due to overripeness).

31. Scientific evidence entered into record for the Waiāhole water case (*In re Waiahole Ditch* 2000). Brasher, Wolff, and Luton (2004) also demonstrate that native aquatic fauna and streams were of higher quality above major diversions than below in Windward Oʻahu streams.

32. Not just forests, but also dew-capturing shrub and grasslands and topography such as high ridges that would harvest the clouds.

33. The heavens are recognized as the source of the purest water (as rain) and the realm where cyclic and seasonal dictates in agriculture originate, such as the planting calendar based on the moon and the *makahiki* season, which marks a "year" and a period of rest for farmers—an ancient festival beginning about October and lasting for four months (Pukui and Elbert 1986).

34. According to Pukui and Elbert (1986, 178), *kula* was a term applied to a plain or open country and later to pasturelands: "An act of 1884 distinguished dry or *kula* land from wet or taro land."

35. At least thirty divisions are described in the literature, some covering broad areas and others very specific ecotypes (Pukui and Elbert 1986; Maly 2003).

36. Relatively few *ahupuaʻa* actually fit the classic description of a pie shaped piece of land from the mountains to the sea. Many were long and narrow, particularly on Maui. *Moku,* larger districts that included many *ahupuaʻa* within their boundaries, more closely resembled the classic form and stretched from the outer reefs to the peaks of the mountains. The *moku* of Waiʻanae traversed from one side of Oʻahu to the other. The eight *moku* of East Maui join at the top of Haleakalā. Trade across *ahupuaʻa* or a chief's responsibility for more than one district made up for the lack of a specific resource within any one land division. A *lele* (detached land parcel) was another mechanism to ensure resource access.

37. A fourth source was *ʻike pāpālua* (deep insight), sometimes inherent from a young age, and sometimes learned.

38. Other types of *kapu* (i.e., social or political) were the purview of higher classes of *kahuna,* such as the *kahuna nui* (high priests) of the chiefs who had mastered a depth of knowledge and multitude of skills befitting the position (Kamakau 1964; Malo 1898).

39. From the viewpoint of a seafaring people in a canoe making landfall or standing in a village in the lowlands, the perspective was ocean to mountain. The *wao akua* (uppermost forest) was the realm of the gods, as was the *pō*, the deepest part of the sea; the coastal lowland, the midpoint inhabited by man.

40. McHarg's precedent setting book of similar title (*Design with Nature,* 1969) urged land use planners to pay attention to the environment and landscapes around them to create more successful urban spaces.

41. While no average field size data is available, individual sugarcane fields in Hawaiʻi are estimated at an average of 300 acres.

42. Observations by the author from the field and archival images, 1987–2013.

43. In the late 1800s, a number was carved in stone at each *loʻi*, possibly to facilitate the assessment of taxes. Stones with numbers exceeding 1,000 have been found.

44. Kamakau (1976, 33) confirms this, citing Campbell's 1819 observation that fields were "seldom above a hundred yards square" unless it were the field of a chief or impor-

tant person. Without additional labor, weeding a single, one-acre field was a monthlong process (taro growers, personal communications).

45. US Department of Agriculture Hawai'i statistics (National Agriculture Statistics Service 2002–2013) show an increase in employment from 1.5 to 2.3 percent and a related decline in unemployment from 9.3 to 8.9 in rural communities between 2009 and 2011. Almost 94 percent of all farms in Hawai'i were smaller than 100 acres in 2007, and the number of fully owned farms climbed from 2,980 (1997) to 5,061 (2007). The number of women as primary farmers doubled in that same period from 922 to 1,806.

46. Kamakau's descriptions (1976, 37) of the large size of *kalo* corms (Ch. 3 endnote 59, p. 72) is possibly a reference to the era before the lifting of the *kapu* in 1819, as he suggests the decline is related to a lack of attention to the appropriate dieties. Each successive era appears to have lost in corm size.

47. Pukui and M. D. (n.d.) allude to this, instructing "of the Lehua taro . . . take the taros with the first grown shoots ('ohā)." Given the evidence that *kalo* was larger (10 lbs or more) in the last two centuries, the *'ohā* (side corms or offshoot) may have weighed several pounds.

48. This can be seen in the *kuaiwi* planting lines of Kailua-Kona, as well. In Kohala these structures may have also served to delineate sociopolitical boundaries (Kagawa and Vitousek 2012; Ladefoged and Graves 2000).

49. Without a Hawaiian language text to compare (none exists today in public archival collections), it is difficult to determine exactly what is meant by "oxidation." While today the term references a browning off or redox of plant material in mucky soils, it appears here to describe a process of letting the soil "breathe."

50. A rare few pictures survive in archival collections of *kalo* growing under ferns, *kukui* forest, and other trees in the uplands.

51. Since the introduction of chemical fertilizers after World War II, the average farmer has added relatively little organic matter to soil banks when compared to historic and ethnographic descriptions.

52. Wetland taro lands are classified as Unique under the Hawai'i IALs designations (Hawaii Revised Statutes § 205-41 through 52). No comparative designation exists for taro lands that lie within the state conservation districts, although legislation has been proposed to remedy this (submitted by the Taro Security and Purity Task Force, Bill SB2241, 2014 Hawaii State Legislature, and HB734 in the 2013 legislative session; http://www.capitol.hawaii.gov/).

53. Kamakau (1976, 37) reasoned that the "afflictions" of the *kalo* (rots, disease, weakness) were a result of a "lax" relationship and lack of *pule* (prayer) to the *'aumākua* (personal gods) of the farmer.

54. Hawaiian time today has come to mean showing up in one's own time, or a bit late.

55. Nupepa Kuokoa, March 24, 1922.

56. Banko et al. (2002) describe the chemical composition and interactions of *māmane* leaves and seeds. A *papa* (decumbent) form of *Sesbania tomentosa* was a known mulch plant in some old *lo'i* sites.

57. Removal of the leaves allowed the plant's energy to move from leaf to root production.

58. A clear example of this is the survival of taro in a variety of niches in Fiji, despite the widespread presence of alomae-bobone virus in some areas (Jackson 2005). That Hawaiian *kalo* varieties continue to survive in the wild on each island is due to the protected topography of the sites, particularly those places out of reach of pigs.

59. While the term "cultivar" is sometimes understood in botany to mean varieties created by selective breeding, among Hawaiian crops, which are clonal in nature, it is not possible to distinguish between land races and deliberate breeding except where recent records or oral history can been found. A plant with desirable traits that can be maintained by propagation and stays true to those characteristics over time is considered a cultivar.

60. Exact numbers are difficult to determine with the loss of cultural knowledge and cultivars that has occurred over the last two centuries. This by no means represents the highest biodiversity for *kalo* cultivars; regions such as Papua New Guinea and Nepal hold this honor (Taro Conservation Strategy Workshop Report 2001).

61. In 1939, sixty-nine Hawaiian taro cultivars were still known to botanical and agricultural collections, along with fifteen Pacific and Asian cultivars; a number of varieties have disappeared over the last thirty years. Fewer than twenty-five *'uala* cultivars are known in local collections; their names and origins remain problematic because of the large number of introductions and hybrids developed in Hawai'i and scant descriptions in written records. Feral cultivars are often found in association with old planting sites. Most remain uncollected and unidentified.

62. Both have recorded brackish water intrusion due to overuse. 'Īao aquifer became the first state designated groundwater management area in 2003; all of Na Wai Ehā came under protection using the same designation process (a contested case) in 2008 (Hawaii Commission on Water Resources Management).

63. Eutrophication and hypoxia of nearshore reefs from agriculture runoff, both signs of excessive nutrient input into those systems, notwithstanding.

64. More detailed discussions of mulch materials and agricultural practices are found in Handy (1940) and in the Hawaiian newspapers, including from such writers as Iokepa, Kalokuokamaile, Kamakau, Luhua, and many unnamed authors.

65. The sap of *kukui* is well known as a treatment for *'ea* or thrush, a fungal infection in children.

66. Recovering native plant species to areas where invasive species are removed brings water cycle recovery full circle. Over time, as invasive species stocks dwindle, managed mulch crops should shift to agricultural lands. On Maui, thousands of acres of invasive bamboo, strawberry guava (*Psidium cattleianum*), rosy apple (*Syzygium jambos*), and grasses await repurposing.

67. Some *kūpuna* feel that animal (or human) wastes were not appropriate for use in the taro patch because of the status of Hāloa. On a practical level, this may have been a cautionary for disease transfer or changing the fragrance or taste of the *kalo*.

68. No data are available on the number of farms using biofuel in Hawai'i.

69. Breadfruit, mango, avocado, and *kalo* are celebrated in festivals enjoyed by residents and visitors in Hawai'i.

70. This calculation is based on a prescribed diet where *kalo* provided one-fourth of starch needs.

71. Feral ungulate (pigs, goats, deer, sheep, cattle) management as a food source has the same challenge as cattle ranching. Herd sizes necessary to meet consumer needs can rapidly push the carrying capacity of the land beyond its limits. Hawai'i's ancient fishponds were an efficient and compact system of protein production.

72. US data relevant to Hawai'i, and for farms of ten acres or less, are difficult to find or disaggregate from sugar, seed, and animal production. The 2012 USDA Census of Agriculture (Hawaii Table 64) finds a total of 4,412 farms of less than ten acres in Hawai'i, 67 percent of which are family or individual owned (3,803); 40 percent of corporate owned farms are also less than ten acres. An estimated 8,368 laborers (hired and unpaid) work on 3,199 farms under ten acres, or an average 2.6 workers per farm. While no survey data exists for taro farms in Hawai'i, an estimated 2–5 workers, often within a single family, are observed on farms of ten acres or less (individual taro growers, personal communication). Small farms are high performers and the basis of economic programs worldwide (e.g., One Acre Fund, 2014). One of the more useful US studies compared labor, inputs, and income for market gardens under three acres, market farms (three to twelve acres), and vegetable farms (greater than twelve acres) and found that while labor inputs (hours) were higher, gross sales and net income per acre were highest in the smallest farms (Hendrickson 2006). In Philadelphia, half-acre urban market farms gross up to $68,000 per year (e.g., Somerton Tanks Farm, n.d.). One UK study reports farms of forty hectares (100 acres) or less provide five times more jobs than those exceeding 200 hectares (500 acres) (Norberg-Hodge, Merrified, and Gorelick 2002).

73. A few countries limit residency by disallowing the ownership of land by foreigners. While this is not likely to be an option for Hawai'i in the near future, there may be alternatives that would allow us to shift *when* someone could buy land, for example, after a person has rented in the islands, contributed to the local economy and learned about where they are living, for a period of time. This would also serve to use the existing large pool of rentals, decreasing speculation and unnecessary development.

74. In 2011, the US Department of Agriculture predicted that the market for locally grown food would reach $7 billion nationally in 2012 (Small Business Labs 2010).

REFERENCES

Andrews, L. [1865] 2003. *A Dictionary of the Hawaiian Language.* With an introduction to the new edition by Noenoe K. Silva and Albert J. Schutz. Honolulu, HI: Island Heritage Publishing.

Banko, P.C., M. L. Cipollini, G. W. Breton, E. Paulk, M. Wink, and I. Izhaki. 2002. "Seed Chemistry of *Sophora chrysophylla* (Mamane) in Relation to Diet of Specialist Avian Seed Predator *Loxioides baileui* (Palila) in Hawaii." *Journal of Chemical Ecology* 28 (7): 1393–1410.

Beckwith, M., ed. and trans. 1951. *The Kumulipo: a Hawaiian Creation Chant.* Honolulu: University of Hawai'i Press.

———. 1985. *Hawaiian Mythology.* Honolulu: University of Hawai'i Press.

Bhongmakapat, T. 2011. *Economics of Enoughness.* Bangkok, Thailand: Chulalonghorn University.

Board of Commissioners of Agriculture and Forestry. 1904. *The Hawaiian Forester and Agriculturalist*, Vol. 1, Nos. 1–12. W. M. Gifford, Honorary Editor. Honolulu: Hawaiian Gazette Publishing Co., Ltd.

Brasher, A., R. Wolff, and C. Luton. 2004. *Associations among Land Use, Habitat Characteristics, and Invertebrate Community Structure in Nine Streams on the Island of Oahu, Hawaii, 1999–2001*. Water Resources Investigation report 03.4256. Reston, VA: US Geological Survey.

Celente, G. 2010. "Neo-survivalism." *Trends Journal* (Winter).

Chun, M. N. 2011. *No Nā Mamo: Traditional and Contemporary Hawaiian Beliefs and Practices*. Honolulu: Curriculum Research and Development Group, College of Education, University of Hawai'i and University of Hawai'i Press.

Decision Analysts Hawai'i, Inc. 2009. *Ma'alaea Mauka Residential Subdivision: Impact on Agriculture. Ma'alaea, Maui*. Draft Environmental Impact Statement, Ohana Kai Village subdivision.

Diamond, J. 2011. *Collapse: How Societies Choose to Fail or Succeed*. Rev. ed. New York: Penguin.

Eliade, Mircea. 1959. *The Sacred and the Profane: The Nature of Religion, the Significance of Religious Myth, Symbolism, and Ritual within Life and Culture*. Translated by W. R. Trask. New York: Harcourt Brace.

Ellis, W. 1825. *Journal of William Ellis: A Narrative of an 1823 Tour through Hawai'i or Owhyhee, with Remarkes on the History, Traditions, Manners, Customs and Language of the Inhabitants of the Sandwich Islands*. Reprint. Honolulu, HI: Mutual Publishing, 2004.

Evans, D., ed. 2008. *Taro Mauka to Makai: Taro Production in Hawai'i*. Honolulu: University of Hawai'i Press.

Fagan, W. F., R. S. Cantrell, and C. Cosner. 1999. "How Habitat Edges Change Species Interaction." *American Naturalist* 153 (2): 165–182.

Föllmi, D., and O. Föllmi. 2003. *Offerings: Buddhist Wisdom for Everyday*. New York: Steward, Tabori and Chang.

Fornander, A. 1916–1919. *Hawaiian Antiquities and Folk-lore: The Hawaiian Account of the Formation of Their Islands and Origin of Their Race, with the Traditions of Their Migrations, etc., as Gathered from Original Sources; with Translations Revised and Illustrated with Notes by Thomas G. Thrum*, Vols. 4–6. Reprint. Memoirs of the Bernice Pauahi Bishop Museum of Polynesian Ethnology and Natural History, 1999. Honolulu, HI: 'Aipōhaku Press.

———. 1969. *An Account of the Polynesian Race: Its Origin and Migrations*. Reprint. Rutland, VT: Charles E. Tuttle.

Gingerich, S. B., C. W. Yeung, T.-J. N. Ibarra, and J. A. Engott. 2007. *Water Use in Wetland Kalo Cultivation*. US Geological Surveys Open File Report 2007-1157. Reston, VA: Office of Hawaiian Affairs.

Gutmanis, J. 1983. *Na Pule Kahiko: Ancient Hawaiian Prayers*. Honolulu, HI: Editions Ltd.

Handy, E. S. C. 1940. *The Hawaiian Planter, Vol. 1, His Plants, Methods and Areas of Cultivation*. Bulletin 161. Honolulu, HI: Bernice P. Bishop Museum.

Handy, E. S. C, and E. G. Handy. 1991. *Native Planters in Old Hawaii: Their Life, Lore, and Environment*. Bulletin 233. Honolulu, HI: Bernice P. Bishop Museum.

Hawaii Revised Statutes § 205-41 through 52. http://www.capitol.hawaii.gov/hrscurrent/.

Hendrickson, J. 2006. *Fresh Market Vegetable Farms at Three Scales of Production*. University of Wisconsin–Madison Integrated Agriculture Systems program.

Hirada, D. 2000. *Local Mythologies, 1979–2000*. http://www2.hawaii.edu/~dennisk/texts/localmythologies.html.

Howes, C., and J. K. Kamakawiwoole Osorio, eds. 2010. *The Value of Hawai'i: Knowing the Past, Shaping the Future*. Honolulu: University of Hawai'i Press.

Hudson, M. 2012. "Hawaii Showcases Its Ag Diversification—the Proof Is in the Numbers." *NASS Blog*, January 17. http://blogs.usda.gov/2012/01/17/hawaii-showcases-it%e2%80%99s-ag-diversification-%e2%80%93-the-proof-is-in-the-numbers/.

In re Waiahole Ditch Combined Contested. 2000. Case Hr'g, 94 Hawaii 97, pp. 112–113.

Iokepa, J. n.d. *Taro Culture*. Collected by Theodore Kelsey. Ms. doc. no. 253. Bishop Museum Archives, Honolulu, HI.

Jackson, G. 2005. *Fact Sheet: Alomae: What Is It?* http://www.ediblearoids.org/Portals/0/TaroPest/LucidKey/TaroPest/Media/Html/Viruses/Alomae/Alomae.htm.

Johnson, R. K. 1981. *Kumulipo: The Hawaiian Hymn of Creation*, Vol. 1. Honolulu, HI: Topgallant.

Johnston, E., and H. Rogers, eds. 2006. *Hawaiian 'Awa: Views of an Ethnobotanical Treasure*. Hilo: Association for Hawaiian 'Awa.

Kagawa, A., and P. Vitousek. 2012. "The *Ahupua'a* of Puanui—a Resource for Understanding Hawaiian Rain-fed Agriculture." *Pacific Science* 66 (22): 161–172.

Kalokuokamaile, Z. P. K. 1922. Ke Ano o ke Kalaiaina. *Nupepa Kuokoa*, May 11 and June 22.

Kamakau, S. K. 1964. *Ka Po'e Kahiko: The People of Old*. Translated by Mary Kawena Pukui. Edited by Dorothy Barrere. Bernice P. Bishop Museum Special Publication 51. Honolulu, HI: Bishop Museum Press.

———. 1976. *The Works of the People of Old: Na Hana a ka Po'e Kahiko*. Translated by Mary Kawena Pukui. Edited by Dorothy Barrere. Bernice P. Bishop Museum Special Publication 61. Honolulu, HI: Bishop Museum Press.

———. 1992. *Ruling Chiefs of Hawaii*. Rev. ed. Honolulu, HI: Kamehameha Schools Press.

Kawaharada, D., ed. 1992. *Hawaiian Fishing Legends with Notes on Ancient Fishing Implements and Practices*. Honolulu, HI: Kalamakū Press.

Kelly, M. 1983. *Na Mala o Kona, Gardens of Kona: A History of Land Use in Kona, Hawai'i*. Bishop Museum Report 83-2. Honolulu, HI: Bernice P. Bishop Museum.

Kepler, A., and F. Rust. 2011. *The World of Bananas in Hawai'i: Then and Now*. Haiku (Maui), HI: Pali-O-Waipi'o Press.

Kirch, P. V. 1994. *The Wet and the Dry*. Chicago: University of Chicago.

Ladefoged, T. N., and M. W. Graves. 2000. "Evolutionary Theory and the Historical Development of Dry-Land Agriculture in North Kohala, Hawai'i." *American Antiquity* 65 (3): 423–448.

Ladefoged, T. N., P. V. Kirch, S. M. Gon III, O. A. Chadwick, A. S. Hartshorn, P. M. Vitousek. 2009. "Opportunities and Constraints for Intensive Agriculture in the Hawaiian Archipelago Prior to European Contact." *Journal of Archaeological Science* 30: 1–10. doi:10.1016/j.jas.2009.06.030.

Laenui, P. 1993. "Indigenous Peoples and Their Relationship to the Environment." Collected by Pōkā Laenui and presented to the Multi-Cultural Center at California State University Sacramento. http://hawaiianperspectives.org/environ.txt.

Leopold, A. 1933. *Game Management.* New York: Scribner's.

Levin, P. 1990. *Farmers Who Forage: Wild Edible Plant Resources, Exploitation and Perception in Southern Thailand.* Paper presented for master's degree, School of Asian and Pacific Studies, University of Hawai'i, Honolulu.

Levin, P., and V. Panyakul. 1993. "Agriculture or Agribusiness? Thai Farmers Search for Viable Alternatives." *ILEIA Magazine* 9 (4): 11–14.

Lukens, A. 2010. "Intersections: The Story of Maui's Four Great Waters." *Green Magazine Hawai'i* 2 (2): 44–51.

Malo, D. 1898. *Moolelo Hawaii: Hawaiian Antiquities.* 2nd ed. Translated by Nathaniel B. Emerson. Bernice P. Bishop Museum Special Publication 2. Honolulu, HI: Bishop Museum Press.

Maly, K. 2003. *Notes on Hawaiian Traditions and Agricultural Practices.* Compiled from a series of oral interviews. Hilo, HI.

McHarg, I. L. 1969. *Design with Nature.* Garden City, NY: American Museum of Natural History, Natural History Press.

Miyasaka, S. 2013. *Conventional Breeding in Taro.* Presentation at a taro field day: "Propagate, Protect, Perpetuate: Waimanalo Taro Collection 2008–2013." University of Hawai'i College of Tropical Agriculture, Honolulu, June 21, 2013.

Mollison, B. 1988. *Permaculture: A Designer's Manual.* Tyalgum, Australia: Tagari Publications.

National Agriculture Statistics Service. 2002–2013. Hawaii [various statistics]. http://www.nass.usda.gov/Statistics_by_State/Hawaii/Search/index.asp.

Norberg-Hodge, H., T. Merrified, and S. Gorelick. 2002. *Bringing the Food Economy Home: Local Alternatives to Global Agribusiness.* Bloomfield, CT: Kumarian Press.

Oahu Resource Conservation and Development Council. 2012. *Sustainable Farming: Master Farmer Series—Paul and Charlie Reppun, Waianu Farm.* Kunia, HI: Oahu Resource Conservation and Development Council.

Oki, D. S., R. H. Wolff, and J. A. Perreault. 2010. *Effects of Surface-Water Diversion on Streamflow, Recharge, Physical Habitat, and Temperature, Nā Wai 'Ehā, Maui, Hawai'i.* Scientific Investigations Report 2010-5011. Reston, VA: US Geological Survey.

Olszewski, D., ed. 2000. *The Māhele and Later in Waipi'o Valley, Hawai'i.* Honolulu, HI: Bishop Museum.

One Acre Fund. 2014. Home page. http://www.oneacrefund.org/.

'Onipa'a Nā Hui Kalo. 2003. *Guidelines for Grassroots Lo'i Kalo Rehabilitation: Pono, Practical Procedures for Lo'i Kalo Restoration.* Printed with the support of Queen Lili'uokalani Children's Center Windward Unit. Punalu'u, HI: 'Onipa'a Nā Hui Kalo.

Parham, J. E., G. R. Higashi, E. K. Lapp, D. G. K. Kuamo'o, R. T. Nishimoto, S. Hau, J. M. Fitzsimons, D. A. Polhemus, and W. S. Devick. 2008. *Atlas of Hawaiian Watersheds and Their Aquatic Resources: Island of Maui.* Honolulu, HI: Bishop Museum, and Department of Land and Natural Resources.

Parsons, R. 2008. "Where's the Water, Brah? East Maui Taro Growers Facing Big Challenges." *Maui Weekly,* December 12.

Preza, D. 2010. *The Empirical Writes Back: Re-Examining Hawaiian Dispossession Resulting from the Mahele of 1948.* Masters thesis. Honolulu: University of Hawai'i.

Prosek, J. 2010. "Maori Eels: New Zealand Maori Defend an Extraordinary Creature—and Themselves." *Orion* (July/August): 14–21.

Pukui, M. K. 1983. *'Ōlelo No'eau: Hawaiian Proverbs and Poetical Sayings.* Bernice P. Bishop Museum Special Publication 71. Honolulu, HI: Bishop Museum Press.

Pukui, M. K., and M. D. n.d. *Taro: Colocasia esculenta (L.) Schott.* Unpublished manuscript. MS Sc Pukui, Box 1-9. Honolulu, HI: Bishop Museum Archives.

Pukui, M. K., and C. Curtis. 1971. *Pikoi.* Honolulu, HI: Kamehameha Schools Press.

Pukui, M. K., and S. H. Elbert. 1986. *Hawaiian Dictionary, Revised and Enlarged Edition.* Honolulu: University of Hawai'i Press.

Pukui, M. K., E. W. Haertig, and C. A. Lee. 1983. *Nānā I Ke Kumu,* Vols. 1 and 2. Honolulu, HI: Hui Hānai Press.

Queen Emma (Emma Kaleleonalani Na'ea Rooke Kamehameha). n.d. [circa mid-1800s]. *Observations on Varieties and Culture of Taro.* Unpublished manuscript. Queen Emma collection 71. Bishop Museum Archives. Honolulu, HI.

Rao, V. R., Hunter, D., Eyzaguirre, P. B., and P. J. Matthews. 2010. "Ethnobotany and global diversity of taro." In *The Global Diversity of Taro Ethnobotany and Conservation,* edited by V. R. Rao, D. Hunter, P. B. Eyzaguirre, and P. J. Matthews, 1–5. Rome: Biodiversity International.

Schumacher, E. F. 1973. *Small Is Beautiful: Economics as If People Mattered.* New York: Harper Torchbooks.

Schwartz, J. D. 2009. "Buying Local: How It Boosts the Economy." *Time Business,* June 11. http://www.time.com/time/business/article/0,8599,1903632,00.html.

Small Business Labs. 2010. *More on the Growth of Small Farms and Local Food.* June 23. http://www.smallbizlabs.com/2010/06/the-continued-growth-of-small-farms.html.

Somerton Tanks Farm. n.d. Home page. http://www.somertontanksfarm.org/about/.

Spoehr, H. 1992. *Kaho'olawe Forest Reserve Period: 1910–1918.* Consultant Report No. 19. Honolulu: Kaho'olawe Island Conveyance Commission.

Sproat, D. K. 2010. "Water." In *The Value of Hawai'i: Knowing the Past, Shaping the Future,* edited by C. Howes and J. K. Osorio, 187–194. Honolulu: University of Hawai'i Press.

Stephenson, W. A. 1968. *Seaweed in Agriculture and Horticulture.* London: Faber & Faber Ltd.

Tadesse, D., and S. C. Blank. 2003. "Cultivar Diversity: A Neglected Risk Management Strategy." *Journal of Agricultural and Resource Economics* 28 (2): 217–232.

Taro Conservation Strategy Workshop Report. 2001. *Taro Genetic Resources: Conservation and Utilization.* TaroGen Project, Suva.

Taro Security and Purity Task Force. 2009. *E ola hou ke kalo; ho'i hou ka 'āina lei'a. The taro lives; abundance returns to the land.* Taro Security and Purity Task Force 2010 Legislative Report. Honolulu: Office of Hawaiian Affairs. http://www. hawaiikalo.org.

———. 2013. Relating to the Protection of Taro Lands, H. 734, 2013 Leg. (Haw 2013). http://www.capitol.hawaii.gov/.

———. 2014. Relating to the Protection of Taro, S. 2241, 2014 Leg. (Haw 2014). http://www.capitol.hawaii.gov/.

Tenney, E. D. 1909. Oahu Water Resources. *Hawaiian Forester and Agriculturist* 6: 131–132.

USDA (US Department of Agriculture) Census of Agriculture. 2012. "Table 64 (Hawaii): Summary by Size of Farm." Washington, DC: US Department of Agriculture. http://www.agcensus.usda.gov/Publications/2012/Full_Report/Volume_1,_Chapter _1_State_Level/Hawaii/st15_1_064_064.pdf.

Watson, L. 1991. *Gifts of Unknown Things: A True Story of Nature, Healing and Initiation from Indonesia's Dancing Island.* Rochester, VT: Destiny Books.

Wianecki, S., and P. Kanahele. 2011. "Pono: Hawaiians Look to the Past to Bring the Future's Forests and Seas into Balance." *Maui No Ka Oi* 15 (2): 29–32.

FIVE

Ecological Design for Island Water Systems

LAUREN C. ROTH VENU

Over 70 percent of the earth's surface is covered with water, yet only a small amount of that water is available for human use: approximately 96 percent of it is salt water, and less than 1 percent is accessible fresh water (Shiklomanov 1993). Over 900 million people around the world lack access to clean water, 2.5 billion people have inadequate sanitation facilities (UNICEF/World Health Organization 2008), and over 3.6 million people die annually from water-related diseases (World Health Organization 2008). These facts are compounded by climate change and corresponding shifts in hydrological cycles. Failing water infrastructures may mean that mounting localized fresh water quality and shortage problems could surpass energy issues in the not-too-distant future (Furumai 2008). According to Charles Fishman (2011, 50), we are "drifting into trouble, failing to get [clean] water to people who don't have it and [doing] an indifferent job of managing water in places that take abundant water for granted."

In Hawai'i and elsewhere, there is a movement to reform human relationships with water and the natural systems that purify it. In turn, this movement encourages us to rethink global, national, and local environmental management and infrastructure development and to protect both the quality and quantity of water resources and the ecological systems that sustain them. Ecological design principles and practices offer techniques that are important for water resource management in the twenty-first century and may offer solutions for

Hawai'i and other isolated communities. This chapter outlines broad problems with existing water infrastructure and access and then gives an example of an ecologically designed system in Hawai'i that harnesses natural processes to help remediate water quality and promote responsible use of fragile Hawaiian water and energy resources.

WATER RESOURCES AND ECOLOGICAL DESIGN

Water resource issues are particularly pronounced in island communities, especially very isolated ones like the Hawaiian Islands. Rainfall is seasonal and plentiful in Hawai'i, but the distribution of water is geographically uneven: rainfall is more plentiful on the windward sides of each island,[1] and the southern and western leeward sides are typically very dry. The ancient Hawaiian land-use pattern of *ahupua'a* reflected this scarcity. It distributed water for multiple uses, including agriculture and human consumption, along an elevation gradient (Cordy 2000; see also chapters 1, 3, and 4 of this volume). The advent of colonial agricultural practices spawned massive engineering efforts to distribute fresh water to dry regions, but the process initiated a pattern of social injustice, which dramatically limited Hawaiians' access to fresh water and limited or terminated many native agricultural practices (see chapters 3 and 4 of this volume).

Disposal or recycling of wastewater is another important consideration for the Hawaiian Islands. Although some islands in the state have invested in significant water-reuse infrastructure, other practices dispose of wastewater into injection wells or the sea,[2] potentially creating greater long-term ecological and financial impacts. To forge a sustainable future for Hawaiian water resources, fresh water availability, water use, and the potential for reuse, the same cycle employed by natural systems must be simultaneously addressed.

Ecological design was defined by Van der Ryn and Cowan (1996, x) as "any form of design that minimizes environmentally destructive impacts by integrating itself with living processes." The cornerstone of ecological design is the wide variety of naturally occurring metabolic pathways that can be used to remediate water-quality issues and other environmental impacts. The methodology looks to nature to identify ways that water infrastructure development can function more sustainably. Ecological design, as an academic discipline, weaves together biology, landscape architecture, planning, and engineering in an attempt to decode nature's adaptive innovations and apply ecological principles to the design of communities, factories, farms, and homes.

Short-term, linear thinking about these infrastructures has led to an illusion that humans are separate from nature. As global human population has increased by a staggering amount,[3] humans are taking more and more resources from the earth. The human ecological footprint has become so large in developed and developing nations that it uses more natural resources than the earth can regenerate each year, including the resources that protect the quality and quantity of fresh water. It is estimated that it takes 1.5 years for the earth to regenerate the resources humans use in one year (Global Footprint Network 2014). This means that by 2050, if things do not change, humans will need more than two Earths to support humanity (World Wildlife Fund 2010). Humans need to address the following question, guided by the principles and science of ecological design: What proactive, adaptive measures can humanity take to benefit the environment such that it can enhance the long-term survival of its species? Rethinking water and wastewater management strategies is key to a new, more sustainable relationship with nature.

IS BIGGER REALLY BETTER? WATER-RESOURCE CONSEQUENCES OF "BUSINESS AS USUAL"

Sustainable development requires that the relationship between human economic systems and ecological systems be such that human life can continue indefinitely without destroying the diversity, complexity, and function of the environment (World Commission on Environment and Development 1987). It occurs "when all humans can live fulfilling lives without degrading the planet" (Global Footprint Network 2010, 37). Alan Weisman (1998) argues that addressing the growing needs of the global human society will require an appropriate balance between nature and technology.

Most nations and business entities currently employ a model that assumes ownership and domination of nature and its resources. "Business as usual" practices in the developed world in the last century have been single-pass design, guided by the westernized mantra that "bigger is better." This model spawned the race to locate and unsustainably harvest natural resources that took millions of years to create. Extraction and burning of fossil fuels, for example, creates significant risks: potential spills, habitat loss, the displacement of wildlife, and the addition of carbon dioxide and other harmful, broader noxious gases into the environment. A hidden cost of fossil-fuel extraction is use of fresh-water resources. Water is intertwined with energy in almost every stage of the process: water is used for extracting petroleum, irrigating biofuels, and mining coal; to refine oil for transportation; and to cool steam in power plants. Energy accounts for one the largest demands of fresh water in the United States: "In 2005, U.S. power plant cooling systems withdrew 143 billion gallons of fresh water per day, accounting for 41 percent of domestic fresh water withdrawals. Mining and fuel extraction withdrew an additional 2 billion gallons per day" (Grubert and Kitasei 2010, 1). Thus, the unsustainable process to extract one resource—fossil fuels—has a negative impact on another vital resource: water.

Water-use and water-availability issues are intertwined with other practices as well. Industrial agriculture draws water from local and regional aquifers, but watering crops hastens the displacement of nutrients from the soil. As a consequence, ever-increasing amounts of chemical fertilizers are added to the ground, contaminating groundwater and streams. Likewise, the growth of impervious surfaces such as asphalt and concrete has repercussions for water resources. Channelized streams diminish water flows, leading to degradation of stream ecosystems and compromising streams' ability to metabolize nutrients effectively (Laws and Roth 2004). Centralized water infrastructure, including centralized potable water and sanitary systems with ocean outfalls, also has significant effects (see chapter 6 of this volume for one example). These practices of plumbing the earth's water systems as part of twentieth-century urbanization have caused, and continue to cause, an alarming, albeit unintended, loss of biodiversity (Nature Conservancy 2008) and negatively affect the survivability of organisms in natural watersheds. Humans are among those organisms. The displacement of fresh water, combined with added pollutants, is creating feedback loops that increase our ecological deficit, and it affects both national and global security by leading humanity into uncharted waters: there is no plan for dealing with this issue economically, socially, or politically.

HISTORY OF CENTRALIZED INFRASTRUCTURE

Modern water infrastructure draws billions of gallons per day out of watersheds, and it has a deleterious effect on the earth's hydrologic cycle on both the supply sources and receiving ends. Ironically, centralized water systems, in particular sanitary systems, were initially designed to protect public health as population densities grew in developed, especially urban areas. The US Clean Water Act of 1972, established to protect public health and the environment, required wastewater to be moved to selected locations for treatment and disposal. However, at the time, wastewater treatment options were primitive, and excess nutrients were often present in the effluents. It was general practice for nutrient-rich effluent to be disposed of through injection wells and open water sources, such as streams, lakes, and coastal areas. This practice was used at Kaneʻohe Bay on the island of Oʻahu, Hawaiʻi, in the 1970s, which received millions of gallons of partially treated water per day. This process shifted the natural ecosystem of the bay, which was once dominated by living corals, to one dominated by algae, sponges, barnacles, and other life-forms indicative of eutrophic conditions (Laws 2000, 99–111).

The displacement of water and nutrients via pipeline can affect the receiving ecosystems at the end of the pipeline and can damage the watersheds from which the water originated. Many pipelines leak, which can lead to water loss or adulteration of groundwater, and large rain events can lead to spills that cause significant damage and pose public health threats. Additionally, rain can move additional fresh water out of the environment as it makes its way into piped infrastructure, rather than refilling aquifers and springs that feed streams and lakes. Pipeline infrastructure has also degraded with age and new infrastructure is needed: Sisolak and Spataro (2005, 5), citing the 2009 American Society of Civil Engineers Report Card, report that "US water and wastewater infrastructure scored a D—with over $255 billion needed to fund upgrades to these systems over the next five years."

There are also considerable hidden energy costs attributable to the inefficiencies of moving vast volumes of water per day from one watershed to another.[4] The California Energy Commission (Klein 2005) conducted a study to evaluate the energy costs of centralized water infrastructure. The study concluded that 19 percent of the state's electricity is used to move and use water, which uses 30 percent of California's natural gas and 88 billion gallons of diesel fuel every year.

The magnitude and rapidity of man-made hydrologic changes, in combination with a growing global population, continue to degrade our water systems and the quantity and quality of available fresh water. As Fishman (2011, 19) states, "Water itself isn't becoming more scarce, it's just simply disappearing from places we are used to finding it." Resource depletion of this kind often leads to social and economic unrest, which historically has led to the growth of resource-controlling tactics amongst banks, utilities, and governments that exacerbate social inequality and unrest. This cycle needs to be broken. "Business as *un*usual" would benefit from an understanding that scale economies of resource depletion may not ultimately be sustainable and will probably have negative impacts on the common good.

THE TWENTY-FIRST CENTURY RENAISSANCE:
ECOLOGICAL DESIGN

The earth's water resources can be used in partnership with nature. Fortunately, solutions are everywhere in natural systems. To address islands' concerns regarding supply and quality of fresh water, the following questions can be asked:

1. What can be learned from other life-forms to effectively harness water in a renewable fashion?
2. What can be learned from food webs so that communities can be designed to better recycle ground nutrients and use water more effectively?
3. How can these systems be used efficiently and positively to stimulate more sustainable, local ecosystems, and economies for the common good?

According to Janine Benyus (1997, 7), a renowned applied ecologist, the field of ecology has collectively laid out the following basic principles and pathways, which move our thinking and practices towards a more sustainable future:

Nature runs on sunlight.
Nature uses only the energy it needs.
Nature fits form to function.
Nature recycles everything.
Nature rewards cooperation.
Nature banks on diversity.
Nature demands local expertise.
Nature curbs excesses within.
Nature taps the power of limits.

These principles are how successful life-forms have sustained themselves over geologic time; people, firms, and government policies can apply them to achieve greater water sustainability. One way of accomplishing this is through biomimicry: a method of mimicking a living process or design in nature that, through the improvement of design efficiency, "lightens" the human footprint on the environment.

ECOLOGICAL DESIGN THROUGH BIOMIMICRY

Biomimicry offers a breakthrough paradigm for humanity to make the transition towards closed-loop, ecologically sound practices for energy, agriculture, building design, resource use, and water management following nature's design strategies. It has been used for designing specific products coined "bionics," the synergy between natural systems and technology. Adaptations found in nature have the potential to offer solutions and efficiencies from which industry can learn. Three examples provide evidence: first, textiles have been strengthened significantly by using threads that have molecular structures similar to spider silk (Benyus 1997); second, solar energy technology attempts to mimic how plants have been harvesting power from the sun for millions of years; and third, the "Bionic" car created by the Mercedes Group, designed after the tropical boxfish to improve both aerodynamics and stability (Block 2005), averages seventy miles per gallon running on a diesel-powered engine (Diseno-Art.com, n.d.). Biomimicry also can be applied to the design of sustainable communities by borrowing from nature the concept of closed-loop processes that minimize waste. Through biomimicry, communities can be designed to create "economies of ecology" that not only are self-sufficient but also add to the stock of natural and economic capital by employing resource recycling that builds rather than depletes resources.

Ecological design is applied biomimicry: it integrates natural systems—in both form and function—into the human landscape to create closed-loop resource management strategies that incorporate lessons learned from natural ecological systems. Biomimicry can address water issues: it can mimic the natural hydrologic cycle in watersheds to maximize the efficiency of fresh water use while also stewarding the ecological system that supports it. Although the term "ecological design" is relatively new, many indigenous cultures worldwide practiced ecological design principles, sustaining large populations for thousands of years. The *ahupuaʻa* concept in precolonial Hawaiʻi, for example, was a social construct defined by the physical watershed in which the community resided, and it highlights principles behind indigenous ecological design.

THE HAWAIIAN *AHUPUAʻA* SYSTEM

In Hawaiʻi, traditional resource management practices were practiced with an ethical and spiritual connection to the land and water through *aloha:* the act of giving back as much (or more) than one receives. Other important Hawaiian values, such as *mālama* (to take care of, and that which nourishes you) and *pono* (equity and righteousness), connected Hawaiians with the *'āina*, or land.[5] Prior to western contact, the last Polynesian influx into the Hawaiian Islands created the *ahupua'a* management scheme, which was locally directed by the resident *ali'i* (chiefs) and *konohiki* (ruler under the chief) and bounded by natural boundaries that typically defined a watershed. Within each of the Hawaiian Islands, districts (*moku*) were formed such that each contained one or more *ahupua'a,* which were physically bounded by the watershed. Moreover, the word "Hawai'i" includes the word water (*wai*), recognizing its essential role for life. Water was central to the *ahupua'a* social construct, whereby the water volume not only determined the number of *ahupua'a* needed per *moku* but also represented the community's relative wealth (referred to as *waiwai*) (Fletcher et al. 2010). There were water and other resource management structures enforced by *kapu,* or restrictions, on what land and water resources could be harvested, when, and by whom. The *kapu* served to protect the purity of water resources. In addition, an integral aspect of the design of the *moku* communities was the reuse and recycling of resources.

Like modern biomimicry practices, Hawaiian agricultural practices were developed through experience with natural hydrologic cycles in order to build resilient, abundant food systems. This resulted in the practice of systematic agriculture, which distinguished the culture of the Hawaiian people from other Polynesian groups (Handy and Handy 1991). Following indigenous ecological principles, the *ahupua'a* system maximized resource and nutrient allocation in such a way that it was able to sustain large populations on the islands for hundreds of years.

NATURAL SYSTEMS TECHNOLOGIES: ECOLOGICAL DESIGN IN PRACTICE

Natural systems technologies (NSTs) are an important approach in ecological design, whereby the knowledge from natural processes can be transferred and applied. NSTs rely on ecosystem development and ecological processes: many operating parts compose the whole, and while working in harmony, they maintain a dynamic equilibrium. Ecological designers study natural systems and processes and apply them to human engineering challenges. For example, innovative solutions for water purification can be modeled on natural wetland ecosystems.

Wetlands perform numerous ecosystem services, including sediment control and nutrient removal (Chescheir et al. 1991), filtering of metals (Keddy 2000), and decomposition of organic matter in the water through various metabolic pathways. These processes support a diversity of plant and animal life while simultaneously filtering, absorbing, and cleaning water. Pioneering German scientist Kathe Seidel undertook seminal research in the 1950s on wetland construction for human-derived wastewater treatment. Seidel examined numerous aquatic plants to determine if any groups of plants could break down or absorb chemical pollutants. Her research revealed that a variety of aquatic plants can perform these cleansing functions. In particular, a common wetland bulrush plant, *Scirpus lacustris,* had the ability to remove phenols, pathogenic bacteria, and other pollutants in wastewater (Seidel, Happel, and Graue 1976). In combination with increased North American interest in wetland functioning, the field of bioremediation, the remediation of pollution using living systems, rapidly grew through the end of the twentieth century. By the mid-1980s, the Tennessee Valley Authority (TVA) led the United States in advancing a full-scale demonstration of constructed wetland remediation projects for a variety of applications, from secondary wastewater treatment to the removal of heavy metals from acid seepage from the mining industry. In turn, TVA hosted the first international conference on constructed wetlands (see proceedings in Hammer 1989). Following this initial conference, the subject still continues to receive significant attention from federal agencies, such as the US Environmental Protection Agency, Soil Conservation Service, and a variety of state environmental agencies. In addition, the NST approach has been applied by engineering firms to treat numerous full-scale wastewater systems throughout the world (Campbell and Ogden 1999).[6] Combining ecological research with newly engineered designs to integrate manmade wetlands into the landscape remains a growing field and represents the application of NSTs in this context.

CASE STUDY: THE HAWAI'I PILOT LIVING MACHINE

Coupled with Hawaiian cultural knowledge of land and water resource management, NSTs may have found an important niche for both onsite wastewater and storm water treatment, embodied in Hawai'i by the Pilot Living Machine (PLM). In February of 2000, the PLM was the first ecologically engineered wetland technology to arrive in Hawai'i; it served as one of the state's first bioremediation pilot projects. The PLM is an ecologically engineered technology pioneered in the 1980s by Dr. John Todd, a renowned ecological designer and founder of Ocean Arks International. Funded by the US Department of Agriculture, the PLM was brought to Oʻahu to evaluate the technology's feasibility on, and applicability to, the islands.

Following the principles of ecological design, Living Machines embody NSTs in that they are specifically designed to emulate ecological principles, including recycling of nutrients, waste, and energy; self-organization; natural treatment processes; and adaptive capabilities. As ecologically engineered systems, these systems harness the natural abilities of living organisms to break down pollutants, metabolize them for energy and nutrients, and self-organize their food web components into robust, contained ecosystems.

The venue for the Hawaiʻi PLM was the Farmer's Livestock Cooperative in Ewa Beach on Oʻahu, a slaughterhouse producing large amounts of wastewater. Prior to the PLM installation, the slaughterhouse was disposing its wastewater in an unlined, open facultative lagoon, which was and continues to be common practice for Hawaiian farmers. The PLM was designed to serve as a secondary treatment system to treat 3,000–6,000 gallons per day of wastewater to meet two water-quality criteria: 10 mg/L total suspended solids and 10 mg/L biochemical oxygen demand. These standards exceed the Hawaiʻi Department of Health R-3 reuse level guidelines.[7] The specific goals of the pilot project were (Todd, Gaddis, and Roth 2002)

1. to demonstrate the treatment performance of a Living Machine in tropical climates, such as Hawaiʻi;
2. to test the performance in wastewater treatment of indigenous and endangered Hawaiian plant species;
3. to test the ability of the system to treat slaughterhouse waste;
4. to use the PLM as a teaching platform for a variety of disciplines, including science, math, management, and cultural history; and
5. to determine the costs of operating the system under steady state conditions over specified time periods.

How the Living Machine Works

The PLM study was designed as a modular unit, constructed out of a steel container. Two separate treatment trains comprised eight tanks each. Each tank was prefabricated, four feet wide and eight feet high. Water levels in each tank were controlled by standpipes located in the last tank for each train. Within each individual tank, several vertical zones were created to provide different physical habitats; each habitat stimulated different biological and chemical reactions. The zones included a well-oxygenated zone at the top of the tank, where floating plant racks were placed; a middle "fixed film" zone with limited oxygen, where media were contained to promote biofilm growth (i.e., biological attachment); and an anaerobic (no oxygen) solids settling and composting zone, to be occasionally flushed and recycled back to the header tanks for further digestion and treatment. An air lift design in each tank circulated water among these zones, moving the water from the bottom to the top before moving to the next tank for further treatment. As the water moved from one tank to the next, it became cleaner; the biological activity digested the organic material coming into the system, which ultimately resulted in a treatment level that exceeded the Hawai'i Department of Health R-3 standards for reuse.

The Role of Plants

Although other Living Machines and wetland technologies on the US mainland sometimes use varieties of tropical plants, the Hawaiian PLM was the first to test native Hawaiian and Polynesian plants for wastewater treatment and bioremediation. Growing native Hawaiian and Polynesian flora was significant for both preserving Hawaiian culture and verifying the use of indigenous Pacific plants in the bioremediation system. Plants were selected from the field and local nurseries with the help of the Bishop Museum, the University of Hawai'i, the US Fish and Wildlife Service, and the Lyon Arboretum to demonstrate and promote the local flora for wastewater treatment, bioremediation, and cultural value.

Vascular plants served several purposes in the PLM system. First, because the plants are typically grown hydroponically in tank-based treatment systems, their roots provided large surface areas to house aquatic microbial communities. Microbial communities hold distinct relationships with plant communities through the exchange of nutrients and enzymes. Second, plants contribute to the release of carbon in water.[8] Lastly, vascular plants take nutrients from the water for growth and reproduction.

The Role of Microorganisms

As in the natural world, microbes are the most abundant organisms in ecologically engineered systems. Utilizing the various zones within the tank complex, a microbial food web developed, consisting of primary producers, consumers, predators, and decomposers. All of the fabricated, engineered components within the PLM were designed to provide habitats for the living components within the system. In turn, the PLM was driven by the microorganisms that stimulated the development of the ecosystems within the PLM, which included multicellular algae, fungi, snails, and fish.

One of the key ingredients to any ecological treatment system is diversity. The biodiversity within the PLM relied upon species diversity, genetic diversity, and bioregional diversity. The PLM was seeded with buckets of active microbial muck, plants, snails, and fish, which were collected with the help of the local US Fish and Wildlife Service from a variety of water environments around the island.

The ecological principles of succession and food web development drive all facets of the Living Machine engines. Waste becomes food for a diverse array of organisms, which have preferences for either the aerobic, anoxic, or anaerobic zones. The design of a robust treatment system relies on the food web developed within the system, which occurs based on the ecological principle of succession. "Succession," as defined in ecology, is the orderly sequence of ecosystem development during which a community of life-forms changes to reach its "climax" state. The first organisms to settle in a community, termed "pioneer species," begin the process of ecosystem development and eventually give way to succeeding organisms that thrive in the conditions created by the pioneers. The initial groups of pioneer organisms in the PLM, principally microbes, take advantage of the raw waste and other physical and chemical conditions in the environment.

To better understand how succession works in the PLM, an analogy can be drawn from how it occurs in nature. For example, on the Hawaiian Islands, lava flows can temporarily destroy most communities, leaving a barren landscape devoid of life. However, shortly afterwards, pioneer species begin to appear, such as soil-making microorganisms and fungi. Soil formation allows seeds to germinate, and subsequently shrubs and grasses emerge. Following their establishment, insects and birds reappear, which feed on and distribute these plant species. Herbivores then become food for other animals. This creates a positive feedback loop, building and expanding upon the primary set of species. While this process evolves, both the physical and chemical components of the environment change, which in turn allows for more plants, microorganisms, fungi, and animals to join the ecosystem. It may require centuries without disturbance for the forests to regenerate, but given enough time, the system will reach its climax state. This process is illustrated by examining the biological evolution of the Hawaiian Island chain as a whole. The older islands, where volcanoes are dormant, have more highly diverse and adapted forests. The ecosystem is both self-regulating and self-sustaining.

In the Oʻahu PLM, under the plants elevated on racks above the water, life in the water column went through a succession process. The only nutrient source available was the slaughterhouse waste stream, delivered from the facultative lagoon, composed of manure and by-products from slaughter. Opportunistic pioneer species that could directly metabolize the waste were initially the most abundant. They, in turn, became a food source, and primary predators began to appear in greater numbers. Next, secondary predators became established in the system. This process continued until the food webs became complex, dynamic, and stable. Observations of microbial and zooplankton diversity development within the PLM indicated that highly developed food webs were formed within a few months, allowing for the development of self-organization within the engineered ecologies. Numerous organisms metabolized nutrients, which cleansed the incoming waste stream to the water quality standard meeting Hawaiʻi Department of Health standards for reuse. The collective organisms' ability to self-organize, capture solar energy, and concentrate nutrients in the food web maximized the biological degradation of contaminants to clean the water.

The PLM program lasted two years. During this time, numerous native Hawaiian, Polynesian, and local plants were tested and rated for the application. In addition, numerous school groups, business leaders, and government officials toured and learned about how wetlands serve the natural world and how they can be applied in the human landscape to treat manmade waste. Several educational curriculum development programs were prepared by faculty at the University of Hawaiʻi at Manoa, and scientific research on the subject continues to be of interest. Shortly after the PLM program ended, the slaughterhouse relocated and expanded. A full-scale natural treatment system unit was engineered and installed at the new location in Campbell, Oʻahu. The PLM found a new home at the Nature Center in Makiki, Oʻahu, where it was renamed the Green Machine and currently treats wastewater from the administration building and reuses it for irrigation. The Green Machine continues to serve as an educational tool for thousands of local K–12 students (Todd, Gaddis, and Roth 2002).

Net zero is a strategy of managing resources within any given property, such that there is little or no reliance on outside inputs. Inherent in this strategy is a decentralized infrastructure that implements best management practices to use and reuse resources locally. The use of distributed resources and decentralized utilities is strategic for communities pursuing sustainability (UN Human Settlements Programme 2009). The concept of net-zero energy has grown throughout Hawaiʻi, and the rest of the United States, through the rapid growth of photovoltaic systems on individual homes and buildings, allowing buildings to generate their own power. Following the success of the renewable energy movement, the concept of net-zero water is also emerging. Net-zero water is the process of collecting, conserving, using, reusing, and infiltrating all water that falls within any given property. It mimics traditional resource management practices in that it maintains water consumption and infiltration within the watershed and the physical property.

Net-zero water models follow a multiuse system design, whereby only potable water needs are met with potable water supply. After its initial use, water can be reused and repurposed for another use on the property. The process starts by evaluating building water needs and applying the three Rs of sustainability as applied to water: reduce, reuse, recycle. First, reduction in water consumption can be achieved by implementing high-efficiency, low-flow fixtures to reduce a building's overall demand for water. Second, the net-zero water methodology identifies potable water needs separately from nonpotable water needs. For example, every home and building needs water, but using nonpotable water sources, such as rainwater or reclaimed water, to flush toilets could save up to an additional 30 percent of potable water. Using rainwater for this purpose is one option; calculating the amount of rainwater that can be collected from the roof and storage capacity during the various seasons will allow the owner to know how much rainwater could supplement nonpotable water needs throughout the year. Stored rainwater could then be filtered and brought back in to the house for use in toilets and other nonpotable water needs as regulators will allow.[9]

Onsite wastewater treatment and reuse are also prerequisites of net-zero water strategies. Wastewater can be classified broadly as either black water or gray water, depending on how it is used. Black water is the untreated water from kitchen sinks and toilets and cannot be reused. Gray water is generated from every other use, including bathroom sinks, showers, and laundry and wash water, and it can account for up to 70 percent of total wastewater from individual homes. It can be reused for some purposes: in areas of Asia and Europe, untreated bathroom sink gray water is reused in the toilet before treatment and final disposal. In the United States, this practice is rare and still in the introductory stage.

Complete onsite wastewater treatment systems (black water and gray water) can be created with wetland technologies, as demonstrated in the PLM example, which produce reclaimed water that meets the State of Hawaiʻi's Department of Health onsite reuse standards. A 1997 US Environmental Protection Agency report concluded that decentralized systems protect public health and the environment, are appropriate for low-density communities and varying site conditions, and they are suitable for ecologically sensitive areas.

For net-zero water buildings to be successfully implemented, there must be a shift in public perception and local permitting regulations, combined with incentives for owners to invest. It requires education and government leadership. Some bright spots are already emerging, however. Examples of government incentives include New York City's Comprehensive Water Reuse Incentive Program, which provides project owners a 25 percent discount on water services for reducing demand on the city's infrastructure for water supply and wastewater services (Sisolak and Spataro 2011). Existing opportunities support the net-zero water challenge, including government funding for Green Reserve projects and, within the private sector, a growing demand for green building accreditation (e.g., LEED and the more progressive Living Building Challenge).[10]

CONCLUSION

Isolated islands worldwide are at a crossroads: continue on the path of "business as usual" or adapt to new methods of development and commerce that are aligned with the laws of nature. Ecological design is a primary alternative methodology for managing water resources with little to no ecological impact. Following the ancient Hawaiian proverb *He aliʻi no ka ʻāina, ke kauwa wale ke kanaka*, The land is the chief, the people are merely its servants (Fletcher et al. 2010), by investing in ecological services and natural capital, it is possible to build economies of ecology that are resilient and equitable, create new technologies based on natural systems, and engineer efficient infrastructure systems that maintain water and nutrients within watersheds by reconnecting humanity back to its natural environment. In the end, humanity has the ability to collectively decide as a society how to forge forward. Fortunately, a global renaissance is upon us, and fortunately all the solutions are abundant in nature—all we need to do is look around.

NOTES

1. Hawai'i comprises eight primary islands: Hawai'i, Maui, Lana'i, Moloka'i, Kaho'olawe, O'ahu, Kaua'i, and Ni'ihau.

2. See chapter 6 in this volume on Maui County's wastewater disposal and water-reuse practices.

3. As of this writing, the population is 7.2 billion and is growing at an accelerating rate.

4. One alternative water infrastructure model would be a decentralized or clustered management and treatment scheme for potable water and wastewater management and reuse, which would make public–private partnerships more feasible and maintain water and nutrients in the watershed.

5. See chapter 1 of this volume for a discussion of the *ahupua'a* system and the cultural practices that sustained precontact Hawaiians.

6. One of the largest constructed wetlands for treating municipal wastewater is in Crowley, Louisiana. It is designed to treat 3.5 million gallons per day (Reed, Crites, and Middle brooks 1995).

7. For a more thorough discussion of the Hawai'i R-3 standard, see chapter 6 of this volume.

8. Carbon is necessary in completing metabolic uptake of nutrients by plants and animals, as defined by the Redfield ratio: the carbon:nitrogen:phosphorus (C:N:P) ratio that individual organisms require for the uptake of any one of these nutrients. Although found to differ between species and groups of species, this ratio is typically 106:16:1 (Redfield, Redfield, Ketchum, and Richards 1963).

9. Precautions to prevent contamination, such as backflow preventers and/or dual plumbing, should be installed when using potable water sourced from the local water utility.

10. For more information on the Living Building Challenge, see https://ilbi.org.

REFERENCES

Benyus, J. 1997. *Biomimicry.* New York: Quill William Morrow.

Block, R. 2005. "The Mercedes Benz Bionic Boxfish Concept Car." June 8. http://www.engadget.com/2005/06/08/the-mercedes-benz-bionic-boxfish-concept-car/.

Campbell, C., and M. Ogden. 1999. *Constructed Wetlands in the Sustainable Landscape.* New York: Wiley.

Chescheir, C. M., J. W. Gilliam, R. W. Skaggs, and R. G. Broadhead. 1991. "Nutrient and Sediment Removal in Forested Wetlands Receiving Pumped Agricultural Drainage Water." *Wetlands* 11: 87–103.

Cordy, R. 2000. *Exalted Sits the Chief: The Ancient History of Hawai'i Island.* Honolulu, HI: Mutual Publishing.

Diseno-Art.com. n.d. "Mercedes-Benz Bionic." http://www.diseno-art.com/encyclopedia/concept_cars/mercedes_bionic.html.

Fishman, C. 2011. *The Big Thirst: The Secret and Turbulent Future of Water*. New York: Free Press.

Fletcher, C., R. Boyd, W. Neal, and V. Tice. 2010. *Living on the Shores of Hawaii*. Honolulu: University of Hawai'i Press.

Furumai, H. 2008. "Rainwater and Reclaimed Wastewater for Sustainable Urban Water Use." *Physics and Chemistry of the Earth* 33: 340–346.

Global Footprint Network. 2010. "Annual Report." http://www.footprintnetwork.org/images/uploads/2010_Annual_Report.pdf.

———. 2014. "Footprint Basics—Overview." http://www.footprintnetwork.org/en/index.php/gfn/page/footprint_basics_overview/.

Grubert, E., and S. Kitasei. 2010. "Briefing Paper 2: How Energy Choices Affect Fresh Water Supplies: A Comparison of U.S. Coal and Natural Gas." Worldwatch Institute. http://www.ourenergypolicy.org/wp-content/uploads/2012/10/natural_gas_BP2_nov2010.pdf.

Hammer, D. 1989. *Constructed Wetlands for Wastewater Treatment*. Chelsea, MI: Lewis.

Handy, C., and E. Handy. 1991. *Native Planters*. Honolulu, HI: Bishop Museum Press.

Keddy, P. 2000. *Wetland Ecology: Principles and Conservation*. Cambridge: Cambridge University Press.

Klein, G. 2005. *California's Water-Energy Relationship*. Sacramento: California Energy Commission. http://www.energy.ca.gov/2005publications/CEC-700-2005-011/CEC-700-2005-011-SF.PDF.

Laws, E. 2000. *Aquatic Pollution*. 3rd ed. New York: Wiley.

Laws, E., and L. Roth. 2004. "Impact of Stream Hardening on Water Quality and Metabolic Characteristics of Waimanalo and Kaneohe Streams, O'ahu, Hawaiian Islands." *Pacific Science* 58: 261–280.

Nature Conservancy. 2008. *Global Impact of Urbanization Threatening World's Biodiversity and Natural Resources*. www.sciencedaily.com/releases/2008/06/080610182856.htm.

Redfield, A. C., B. H. Ketchum, and F. A. Richards. 1963. "The Influence of Organisms on the Composition of Sea-Water." In *The Composition of Sea-Water*, edited by M. N. Hill, 2: 26–77. New York: John Wiley & Sons, Inc.

Reed, S., R. Crites, and E. Middlebrooks. 1995. *Natural Systems for Waste Management and Treatment*. 2nd ed. New York: McGraw-Hill.

Seidel, K., H. Happel, and G. Graue. 1976. *Contributions to Revitalisation of Waters*. Krefeld-Hulserberg: Limnologische Arbeitsgruppe in der Max Planck Gesellschaft.

Shiklomanov, I. 1993. "World Fresh Water Resources." In *Water in Crisis: A Guide to the World's Fresh Water Resources*, edited by P. Gleick, 13–23. New York: Oxford University Press.

Sisolak, J., and K. Spataro. 2011. *Toward Net Zero Water: Best Management Practices for Decentralized Sourcing and Treatment*. Portland, OR: Cascadia Green Building Council. http://www.ecobuildingpulse.com/Images/TNZW_tcm131-1075029.PDF.

Todd, J. H., E. Gaddis, and L. Roth. 2002. Final Report on the Slaughterhouse Pilot Living Machine. Report to the US Department of Agriculture, Washington, DC.

UN Human Settlements Programme. 2009. *Planning Sustainable Cities: Policy Directions: Global Report on Human Settlements 2009,* Earthscan. http://www.rrojasdatabank.info /plansustcities1.pdf.

UNICEF/World Health Organization. 2008. *Progress on Drinking Water and Sanitation: Special Focus on Sanitation.* http://www.unicef.org/media/files/Joint_Monitoring _Report_-_17_July_2008.pdf.

US Environmental Protection Agency. 1997. *Response to Congress on Use of Decentralized Wastewater Treatment Systems.* EPA 832-R-97-001b. Washington, DC: US Environmental Protection Agency.

Van der Ryn, S., and S. Cowan. 1996. *Ecological Design.* Washington, DC: Island Press.

Weisman, A. 1998. *Gaviotas: A Village to Reinvent the World.* White River Junction, VT: Chelsea Green Publishing Company.

World Commission on Environment and Development. 1987. *Our Common Future.* Oxford: Oxford University Press.

World Health Organization. 2008. "Safer Water, Better Health." http://www.who.int/quan tifying_ehimpacts/publications/saferwater/en/index.html.

World Wildlife Fund. 2010. *Living Planet Report 2010: Biodiversity, Biocapacity and Development.* http://www.footprintnetwork.org/press/LPR2010.pdf.

SIX

Saving Island Water

Strategies for Water Reuse

STEVE PARABICOLI

In Hawai'i, Maui County's Wastewater Reclamation Division (WWRD) is recognized as a water-reuse leader. The objectives of Maui's water-reuse program
are to supplement Maui's limited potable water supply and to reduce the use
of injection wells for effluent disposal. Recycled water is reused from all five
of the county's wastewater reclamation facilities (WWRFs), and significant distribution systems have been constructed in South and West Maui. Recycled
water is now used for a wide variety of purposes, including landscape irrigation, agricultural irrigation, cooling, fire control, composting, toilet flushing,
environmental enhancement, and construction purposes. Key components that
contributed to the success of the WWRD's reuse program include water-
reuse feasibility studies, the creation of a water-recycling program coordinator
position, WWRF upgrades, recycled water distribution system development,
passage of a mandatory recycled water use ordinance, the development of a
community-based rate structure, and proactive public education.

The initial primary factor driving Maui's water-reuse program was public
concern: wastewater disposal in injection wells was suspected to be the main
cause of periodic seaweed blooms in Maui's coastal waters. While this concern
still lingers, increased development and the lack of available fresh water has
also made recycled water a water-resource issue. Recycled water is a valuable

resource in Maui County, and as a result, developers are funding expansions to Maui County's recycled water distribution systems. These expansions benefit Maui's community and environment through potable water savings and reduced effluent disposal in injection wells.

MAUI COUNTY'S WATER-REUSE EXPERIENCE

The County of Maui consists of three populated islands: Maui, Moloka'i, and Lana'i. Maui County's WWRD is responsible for the collection and treatment of the majority of wastewater that is generated. The WWRD operates three wastewater systems on Maui, one on Lana'i, and one on Moloka'i (fig. 6.1). The Maui island facilities utilize activated sludge with biological nutrient removal, sand filtration, and disinfection with either ultraviolet (UV) light or chlorine. Twenty-six small private wastewater treatment plants (twenty-two on Maui, three on Moloka'i, and one on Lana'i) are located in resort areas and regions that are not serviced by the county's wastewater system. There are also some WWRD individual wastewater systems at residences.

FIGURE 6.1. Maui County, with locations of county wastewater reclamation facilities.

Maui County's WWRFs process 12 million gallons per day (mgd) of waste-water. Table 6.1 defines types of recycled water; R-1 (tertiary) recycled water is produced at the South Maui and the West Maui WWRFs, while R-2 (disinfected secondary) recycled water is produced at the Central Maui WWRF. Each Maui island facility currently treats approximately 4.0 mgd. The WWRFs on Lanaʻi and Molokaʻi are much smaller. The Lanaʻi WWRF treats 0.28 mgd to R-3 (undisinfected secondary) quality using stabilization ponds. All of the effluent is then sent to a private auxiliary facility and upgraded to R-1 quality

TABLE 6.1. Recycled water definitions

Water Type	Permitted Uses
R-1 *water* is tertiary treated recycled water, the most highly treated type produced. It has undergone a significant reduction in viral and bacterial pathogens. As defined by the guidelines, R-1 water is recycled water that is at all times oxidized, then filtered, and then exposed, after the filtration process, to disinfection processes that remove 99.99 percent of some viruses, and shows extremely low levels of fecal *E. coli* bacteria.	R-1 water can be utilized for spray irrigation without restrictions on use. It is now approved for applications including spray irrigation of golf courses, parks, athletic fields, school yards, residential properties where managed by an irrigation supervisor, and roadsides/medians; for drinking water by nondairy animals; and for vegetables and fruits that are eaten raw.
R-2 *water* is disinfected secondary treated recycled water; it is treated, but not as much as R-1 water. As defined by the guidelines, R-2 water means recycled water that has been oxidized and disinfected to meet low, but not extremely low, *E. coli* concentrations.	When using R-2 water, spray irrigation is limited to evening hours, and a 500-foot buffer zone between the approved use area and adjacent properties is required. Several golf courses in Hawaiʻi are irrigated with R-2 water. Food crops that are irrigated with R-2 water either must be irrigated via a subsurface irrigation system or, if irrigated with spray irrigation, must undergo extensive commercial, physical, or chemical processing determined by the Hawaiʻi Department of Health to be sufficient to render it free of viable pathogenic agents, before it is suitable for human consumption.
R-3 *water* is undisinfected secondary treated recycled water, and there are severe limitations on its use.	R-3 water can be used via surface, drip, or subsurface irrigation of pastures, seed crops not eaten by humans, orchards, and vineyards, where the recycled water does not come in contact with the edible portion of the crop; and it can be used for food crops that undergo extensive commercial, physical, or chemical processing, determined by Hawaiʻi Department of Health to be sufficient to render it free of viable pathogenic agents, before it is suitable for human consumption.

Source: Hawaiʻi Department of Health (2002).
Note: The Hawaiʻi Department of Health does not permit any of the classes of recycled water defined here to be used for potable purposes by humans.

using water hyacinths, coagulation, sand filtration, and UV disinfection. The WWRF on Moloka'i treats 0.18 mgd to R-2 quality using rotating biological contactors, sand filtration, and chlorine disinfection.

As of 2011, approximately 3.3 mgd, or 27 percent, of the recycled water produced at Maui County's WWRFs was reused. Most of the water-reuse projects on Maui are located in South and West Maui and are provided with R-1 water. On the island of Lana'i, R-1 water is used to irrigate a large resort golf course. On the island of Moloka'i, approximately 1,000 gallons per day of the R-2 recycled water is utilized for the irrigation of native Hawaiian plants along the main highway.

Recycled water from WWRFs that is not reused is disposed into injection wells on Maui. The effluent (~8.6 mgd) eventually reaches the ocean. There has been community concern that the nutrients in wastewater effluent may be contributing to periodic nearshore seaweed blooms and coral reef decline, which played an important role in the development of Maui County's water-reuse program.

PROGRAM DEVELOPMENT

In the early 1990s, Maui County's WWRD developed a plan and embarked on a long-term program to reuse millions of gallons of high-quality recycled water that had been historically disposed of in injection wells. The driving factor that initiated the development of Maui's water-reuse program was related to effluent disposal. The US Environmental Protection Agency (EPA) and local environmental groups expressed a concern that injection wells may contribute nutrients that cause algae blooms in coastal waters. A West Maui watershed committee was formed in the mid-1990s to investigate the cause of the algal blooms in West Maui. Several studies were conducted as a part of this effort. No conclusive evidence was found that linked the injection wells at the WWRF to the algal blooms. Nevertheless, in 1995, the US EPA placed a 6.7 mgd limit of effluent and a total nitrogen limit on effluent of ten milligrams per liter that could be disposed of in the injection wells at the West Maui WWRF.

Researchers from the US Geological Survey at the University of Hawai'i conducted further studies on Maui in 2006. They identified the location of the WWRF effluent plume and the presence of the isotopic form of nitrogen ($\delta^{15}N$) typically found in human wastewater in seaweed in the nearshore waters on the south side of Maui (Hunt 2006). Dailer et al. (2010) conducted a Maui island-wide coastal survey using $\delta^{15}N$ as an indicator to demonstrate that the highest algae $\delta^{15}N$ levels were found in locations close to Maui County's three largest wastewater treatment facilities. The newly documented effect of effluent disposal raised further public concern that injection wells were degrading Maui's coastal environment. In addition to effluent disposal, water supply was another important factor that led to the development of the reuse program. Increased development and the lack of available fresh water made recycled water an attractive nonpotable water option in certain areas of Maui. As a result of these issues, Maui County decided to expand its water-reuse program. Several key components were identified that formed the foundation of the program.

Water-Reuse Feasibility Studies

The WWRD hired a consultant to conduct water-reuse feasibility studies in 1990–1991 for the three main populated areas of the island of Maui served by county WWRFs (South Maui, West Maui, and Central Maui). It was determined that the two areas that would benefit the most from the development of recycled water distribution systems were South and West Maui. Both areas are relatively arid and do not have adequate supplies of fresh water resources. Nonpotable brackish groundwater quality is also typically poor in these areas because of salinity content. The feasibility studies identified a number of potential properties that could utilize recycled water for irrigation, as well as potential routes for recycled water distribution systems.

Water-Recycling Program Coordinator Position

In 1993, Maui County became the first county in Hawai'i to create a water-recycling coordinator position. This was an important step because a primary responsibility of this position was to champion the cause of water reuse in the county. For the first time, a full-time position was dedicated to promoting and implementing water-reuse initiatives. The responsibilities of the position included short- and long-range planning, conducting proactive public education, interacting with water-reuse customers and assisting them with all phases of recycled water project management, inspecting new and existing recycled water irrigation systems, seeking potential uses for recycled water, and researching potential funding options for system expansions. Day-to-day operations and maintenance of recycled water distribution systems also fell under the responsibility of this position.

WWRF Facility Upgrades

Based on the findings of the feasibility studies, the county upgraded South and West Maui's WWRFs to R-1 quality in 1993. Both facilities added turbidity monitoring, coagulation capability, effluent filtration, and UV disinfection to meet the R-1 recycled water requirements, as required by the Hawai'i Department of Health *Guidelines for the Treatment and Use of Recycled Water* (2002). UV capacity was installed at the South Maui WWRF to disinfect the entire design flow of 8.0 mgd to R-1 quality. One UV channel with 2.0 mgd capacity was installed at the West Maui WWRF.

Recycled Water Distribution System Construction

In 1995–1996, the county constructed recycled water distribution systems in South and West Maui. General obligation bonds and state revolving funds with low-interest loans were used to fund the construction of these systems. The South Maui system was designed with a covered, elevated storage tank that provides on-demand pressure for water-reuse projects and recycled water to several commercial properties in the area.

Recycled Water Mandatory Use Ordinance

Public concern prompted a 1995 US EPA requirement that limited the daily volume of effluent disposed of into injection wells. Since the facility's daily flow was already approaching this volume, increasing the volume of recycled water used from the facility helped meet the requirement. Maui County became the first county in Hawai'i to pass an ordinance that required commercial properties to utilize recycled water for irrigation purposes. The ordinance required all improved commercial properties that are within one hundred feet of an R-1 water distribution system to link to the system within one year of recycled water availability (Maui County Code 1996). The ordinance established three main recycled water customer classes: (1) major agriculture: agricultural consumers that use more than three million gallons of recycled water per day, or that have more than 400 acres served by the recycled water distribution system, or that have any pastureland served by the recycled water distribution system; (2) agriculture: agricultural consumers, including golf courses that use less than three million gallons of recycled water per day; and (3) all others, such as parks, schools, condominiums, hotels, shopping centers, construction, and industrial uses. The ordinance was a significant factor in convincing Maui County's largest recycled water consumer, a West Maui resort, to switch its irrigation system from brackish groundwater to recycled water. The ordinance was adopted as part of the Maui County code. To support this legislation, rules and regulations for recycled water service were adopted shortly after the passage of the ordinance.

Community-Based Recycled Water Rate Study

In 1995, a consultant was hired to assist with the development of the recycled water rate structure. Subsequently, a committee was formed to provide input and evaluate the various rate scenarios submitted to the WWRD by the consultant. The committee consisted of representatives from the WWRD, the County of Maui's Department of Finance, private large land and golf course owners, the Maui Hotel and Lodging Association, the Chamber of Commerce, real estate agent groups, and the consultant. Committee members agreed that recycled water users should not endure the burden of financing the entire recycled water program and that the sewer users should share in the cost of "disposing" of recycled water through water reuse. Another important goal established by the committee was to set recycled rates slightly less than the rates paid for other traditional water sources such as brackish groundwater, stream water, or potable water. This provided a financial incentive for consumers to convert irrigation systems to recycled water.

The result was a composite rate structure that set rates at reasonable levels and recovered expenditures for recycled water distribution infrastructure construction, operations, and maintenance from both recycled water users and sewer users. Under the new structure, sewer users paid for 75 percent and recycled water users paid 25 percent of the recycled water program costs. Recycled water rates were set for each of the three main customer classes identified in the mandatory use ordinance and were set slightly below the costs of conventional water sources. Connection and meter fees for South and West Maui were established at this time (Parabicoli 2001). To jump-start the program, connection fees were waived for the first two years. A later addition to the rate structure was an Avoided Cost customer classification, which allows the county to match the rate that consumers previously expended on their nonpotable water source.

Proactive Public Education

As part of the WWRD's efforts to set the water-reuse program on a strong foundation, an aggressive public outreach program was launched to proactively educate politicians (administrative and legislative), the community, and potential recycled water users about the benefits of, and safety aspects associated with, recycled water use. The WWRD learned from other municipalities throughout the country that failure to include this important step in their program could lead to substantial delays and even cancellations of water-reuse projects due to a lack of understanding. Politicians were targeted early on because their support was needed to procure funding for construction of recycled water infrastructure and to pass the mandatory recycled water use ordinance. The WWRD was successful in convincing local government leaders to approve legislation and funding that laid the foundation for Maui's water-reuse program.

The public education program has continued and has been expanded to all segments of the community. A variety of techniques have been utilized to educate the public, including presentations to county boards and commissions, community groups, senior citizen groups, engineering and landscape architect groups, schools, and recycled water users. WWRF tours, participation in environmental awareness events, periodic press releases announcing the start-up of new recycled water projects, and the distribution of promotional items (bumper stickers, magnets, rulers, etc.) were used to increase public awareness. Open and honest communication with the public resulted in widespread acceptance of the WWRD's water-reuse program.

WATER-REUSE PROGRAM RESULTS

Maui County's WWRD reused 27 percent (one billion gallons or 3.3 mgd) of the wastewater processed at its WWRF in 2010. Most of the recycled water is used for landscape irrigation. Significant volumes are also used for industrial purposes (mainly at county WWRFs) and agricultural irrigation. Landscape irrigation occurs at golf courses, parks, schools, multifamily residential complexes, a community center, a public library, a fire station, a shopping center, roadside shoulders and medians, and common areas. Other innovative uses of recycled water exist, such as green waste composting, toilet/urinal flushing, construction purposes, environmental enhancement (soil erosion control and wildlife habitat), and fire control. Table 6.2

TABLE 6.2. Year 2010 recycled water use categories and volumes used

Category	Volume Used (million gallons)	% Total Use
Landscape irrigation	934.41	77.75%
Industrial use	144.86	12.05%
Agricultural irrigation	53.37	4.44%
Construction purposes	41.50	3.45%
Composting	3.48	0.29%
Environmental enhancement	24.27	2.02%

lists the various uses of recycled water and volumes/percentages for the year 2010.

The majority of the water-reuse projects are located in South Maui because the system has sufficient R-1 recycled water production capability and storage capacity. Elevated storage allows the distribution system to be continuously pressurized and reduces recycled water pumping requirements. As a result, commercial properties that are connected to the system have 24-hour per day recycled water service. The WWRF also has a 1.8 million gallon recycled water storage reservoir onsite. The reservoir and the elevated 1.0 million gallon tank are covered to maintain water quality. The WWRD funded most of the infrastructure through a state revolving fund loan; however, in 2006, developers funded an extension of the system that allowed them to utilize the R-1 recycled water for landscape irrigation.

SUCCESSFUL WATER-REUSE EXAMPLES IN MAUI COUNTY

Example 1

A local golf club adjacent to the South Maui WWRF has successfully utilized recycled water from the facility since the course was constructed in 1986. The golf course was built with the intention of using recycled water as its sole source of irrigation water. Up to 1.0 mgd of R-1 recycled water is utilized. The golf course utilizes pressure from the county's R-1 recycled water distribution system to operate its spray irrigation system. Since the WWRF incorporated biological nutrient removal into its secondary treatment process, the nutrient value of the R-1 recycled water decreased and the golf course required more supplemental fertilization. The lower nutrient value, however, contributed to less algal buildup in water hazards that contain the recycled water.

Example 2

Another local company utilizes microalgal technology to produce the antioxidant astaxanthin, an ingredient used in the health food industry. The company was attracted to South Maui due to its abundant sunshine and the availability of R-1 recycled water. Since 2000, it has used 0.02 mgd of R-1 water from the WWRF for cooling the biodomes in which algae are grown. A curtain of recycled water maintains optimum temperatures for the algae. After the biodomes are cooled, the recycled water is collected and sent back to underground storage, where it was cooled and pumped back to the biodomes. The company's facility also utilizes R-1 recycled water for fire control. This project is an example of how recycled water has stimulated economic development on Maui.

Example 3

A West Maui resort has been using up to 1.3 mgd of R-1 recycled water since 1997, making it the WWRD's largest volumetric recycled water user. Prior to utilizing recycled water, high-salinity brackish water was used as the resort's sole irrigation water source. The county's mandatory recycled water use ordinance was influential in convincing the resort to convert to recycled water irrigation. Recycled water is pumped from the WWRF directly to an open reservoir located at the resort. The R-1 water is used to irrigate the lower portions of two golf courses, road shoulders along the main highway, and median strips within the resort. Since the recycled water is lower in salinity than the traditional brackish groundwater that was previously used, the appearance of the turfgrass at the resort has improved significantly since recycled water use commenced in 1997. The resort qualified as an Avoided Cost customer since it previously used nonpotable water for irrigation. The WWRD matched the rate that was paid by the resort for its brackish water supply.

OTHER NOTABLE USES OF RECYCLED WATER IN MAUI COUNTY

Dust Control by Contractors

Because of potable water shortages in Maui County, the Department of Water Supply has restricted the use of potable water for dust control and other construction activities. Construction companies are now required to use nonpotable water sources for these activities. Because of the restriction on using potable water, increasing amounts of recycled water are being used by more construction companies. The WWRD has constructed recycled water fill stations at its three Maui island facilities to allow contractors to fill water tanker trucks. Recycled water fire hydrants are also utilized as fill locations in South Maui.

Island of Lana'i

A world-class golf course located on the island of Lana'i opened to the public in 1991. The course was irrigated with potable water, but the county ordinance required the course to switch to nonpotable water within five years. The owners of the golf course constructed the Lana'i City Auxiliary Wastewater Treatment Facility in 1994. The facility receives all of the daily flow (0.25 mgd) of secondary treated effluent from the county's stabilization pond treatment plant. R-1 water is produced at the auxiliary facility through the use of water hyacinths, chemical coagulant addition, up-flow sand filtration, and UV disinfection.

R-1 recycled water is the primary irrigation source for the golf course. It is also used in the water features at the course. Because the daily volume of 0.25 mgd is a relatively small volume of water to irrigate an eighteen-hole golf course, maintenance personnel employ extreme water conservation measures to extend the recycled water supply. Rainwater is channeled to irrigation ponds located on the course. The course has been planted with drought-tolerant Bermuda grass, and weather stations are situated throughout the course so that watering takes place only when it is necessary. Soil moisture content is also measured on a regular basis. Because the course is located over a potable water aquifer, a groundwater monitoring program required by the Hawai'i Department of Health's 1993 water-reuse guidelines has been in place since recycled water use was initiated in 1994.

CURRENT CHALLENGES

Maui County's WWRD made great strides in the 1990s to create a strong foundation for its water-reuse program. Key legislation and the construction of R-1 water production and distribution infrastructure were accomplished during this period. Many projects with innovative uses of recycled water have since been initiated, and Maui is recognized throughout Hawai'i as a leader in water reuse. The WWRD received an award in 2007 for its sustainable practices from *Hawai'i Home and Remodeling* magazine's Who's Keeping Hawai'i Green?

environmental awards program. Unfortunately, much of the funding for the WWRD has been transferred to sewage projects. In 1999, the US EPA issued Maui County a consent decree due to the occurrence of sewage spills from its wastewater collection system that took place in previous years. As a result, sewer improvement projects, such as the replacement of aging sewage force mains or adding capacity at WWRFs and pump stations, has taken precedence over the addition of more recycled water production and distribution infrastructure. Sewer user fees have been significantly increased to pay for these required improvements and are expected to increase further as more of these improvements are constructed. Subsequently, WWRD approval of capital improvement projects that were supposed to expand to both South and West Maui recycled water distribution systems was delayed. A West Maui water-reuse system expansion project that was to receive a grant from the US Bureau of Reclamation Title XVI program was also postponed because the WWRD did not have the required matching funds.

The concerns that injection wells are detrimental to the coastal environment near the county's Kīhei WWRF prompted the Pukoʻa O Kama Alliance, a group of South Maui residents, to file a lawsuit against the County of Maui in 2009. Around this same time period, a group of concerned citizens formed a coalition named DIRE, for "Don't Inject, Redirect." This coalition, along with other local environmental groups, urged the county to place a higher priority on its water-reuse program. The EPA, reacting to the concern of Maui's environmental groups regarding injection wells, proposed strict limitations on the nitrogen content of the injected effluent at the West Maui WWRF and required that all injected effluent from the facility be disinfected to R-1 recycled water standards.

The recycled water production and distribution infrastructure that was constructed by the county was funded through bonds or state revolving fund loans. The WWRD incurred all of the debt service associated with the construction of the water-reuse system. Operation and maintenance of the recycled water production and distribution systems are also paid for by the WWRD. Despite the fact that over 400 million gallons of potable water annually are being saved through WWRD's water-reuse program, the County of Maui's Department of Water Supply has not contributed funding towards recycled water distribution system development. Thus, the WWRD, a municipal wastewater agency, has taken on the role of a water purveyor in Maui County. The WWRD has, for many years, operated and maintained wastewater collection and treatment systems. However, the WWRD has little experience as a water purveyor. The recycled water distribution infrastructure constructed by the WWRD in the 1990s, including storage tanks, valves, pressure reducing systems, and metering systems, requires maintenance, and the WWRD overlooked this requirement. To address this challenge, the water-recycling program coordinator has used a combination of tactics to ensure that the recycled water distribution system is properly maintained.

As recycled water systems are expanded on Maui, it may be necessary to create a "mini" water department within the WWRD. This section will have certified water distribution operators and maintenance personnel whose main responsibility will be to operate and proactively maintain the recycled water distribution systems.

Having the WWRD as the water purveyor is quite a different scenario than what has taken place on the island of Oʻahu, where the city and county's Board of Water Supply has assumed responsibility of the production and distribution of recycled water. They have recognized the value of recycled water and have been very progressive in developing it into a viable nonpotable water option on the island. (Parabicoli and Brown 2004). Maui County's Department of Water Supply is, however, including the use of recycled water in its water-use and development plans and may contribute to the development of Maui's recycled water distribution system in the future. Despite the funding challenges, the County of Maui's recycled water distribution system is slowly growing. Because of potable water shortages on Maui, developers are now funding expansions to the South and West Maui recycled water distribution systems.

LESSONS LEARNED FROM MAUI'S WATER-REUSE EXPERIENCE

Maintain the Momentum

The WWRD set up a water-reuse program with a strong foundation and started constructing infrastructure to reuse recycled water and reduce the use of injection wells. As time passed, funding from the WWRD for the continued growth of the water-reuse program was discontinued due to higher-priority sewer system rehabilitation projects. It is difficult to speculate on whether legal action from environmental groups, and the imposition of strict injection well discharge requirements at the West Maui WWRF from the US EPA, could have been avoided if Maui County had maintained the steady growth of its recycled water distribution systems, thus increasing the use of recycled water and reducing the use of its injection wells. It could be argued, however, that by maintaining a steady, proactive commitment to water reuse, the system may have been further along.

Share the Financial Burden

The WWRD took on the financial challenge of developing recycled water infrastructure, and as a result, sewer system users pay for the majority of the water-reuse program costs. Since the use of recycled water can save potable water and offer many other environmental benefits, it is recommended that funding for the development of recycled water production and distribution infrastructure be obtained from several sources, including sewer users, potable water users, property taxes, the visitor industry, developers, and state agencies. Grant opportunities, while limited, should also be pursued when they are available.

Involve the Community

Open and honest communication with the community is always required when developing a water-reuse program. The WWRD involved the community during the early stages of the water-reuse program and maintains its public education outreach. This consistent effort has led to widespread acceptance of the concept of using recycled water at schools, parks, and commercial properties frequented by the public.

Create a Dedicated Operation and Maintenance Staff

The WWRD constructed recycled water distribution systems but did not assign staff to operate and maintain the infrastructure. If a wastewater utility takes on the role of a recycled water purveyor, staff should be trained and dedicated to operate and maintain recycled water distribution systems. If this is not possible, the operation of recycled water systems should be turned over to municipal water departments, because such agencies already have experience in operating and maintaining water distribution systems.

Assign a Champion

The creation of the water-recycling program coordinator position within the WWRD was important because this full-time position was dedicated to championing water-reuse initiatives. Any municipality that is starting a water-reuse program should assign a program coordinator rather than "piecemeal" the responsibilities of program management to other staff, whose responsibilities to the water-reuse program will compete with other tasks.

Elevate the Priority

Water reuse contributes to sustainable water resource management, and recycled water is a drought-proof water resource. The increased use of recycled water also reduces effluent disposal requirements. Thus, water reuse addresses community concerns that effluent discharges are detrimental to the environment. For these reasons, municipalities facing water-resource and wastewater disposal challenges should place a high priority on developing recycled water production and distribution capabilities.

CONCLUSION

The County of Maui's WWRD has created a water-reuse program that has a strong foundation, and several properties now utilize recycled water for a variety of purposes. The use of recycled water in Maui County has resulted in potable water savings and reduced the volume of effluent disposed of into injection wells that eventually reach the ocean. In many cases, businesses have been drawn to Maui due to the availability of recycled water, and economic development has been stimulated. Despite the success of the WWRD's water-reuse program, recycled water is still underutilized in Maui County. A significant financial commitment is required to develop the necessary production and distribution infrastructure to increase recycled water use. However, the value of recycled water is increasing as continued development stresses available freshwater resources. The concern that injection wells, typically utilized for effluent disposal, are contributing to periodic algal blooms and coral reef decline in Maui's nearshore coastal waters continues to be a significant driving factor for water reuse on the island. This factor, litigation from environmental groups, and regulatory requirements from the EPA will require the WWRD to once again elevate the priority of expanding its water-reuse program. The innovative uses of recycled water discussed in this chapter exemplify the point that recycled water can further contribute to sustainable water-resource and environmental management for island communities.

REFERENCES

County of Maui. 2011. "Use of Reclaimed Water. Maui Index, 1996, 1997, 1998, Chapter 20.30." http://ordlink.com/does/maui/index.htm.

Dailer, M. L., R. S. Knox, J. E. Smith, M. Napier, and C. M. Smith. 2010. "Using $\delta^{15}N$ Values in Algal Tissue to Map Locations and Potential Sources of Anthropogenic Nutrient Inputs on the Island of Maui, Hawai'i, USA." *Marine Pollution Bulletin* 60: 655–671.

Hawai'i Department of Health. 2002. *Guidelines for the Treatment and Use of Recycled Water*, rev. Hawai'i Administrative Rules, chap. 11–62, 12–13. Honolulu: Hawai'i State Department of Health.

Hunt, C. D. 2006. *Ground-Water Nutrient Flux to Coastal Water and Numerical Simulation of Wastewater Injection at Kihei, Maui, Hawai'i*. Scientific Investigations Report 2006-5283. Honolulu, HI: US Department of Interior, US Geological Survey.

Maui County Code of Ordinances. 1996. Title 20 —Environmental Protection, Chapter 20.30—Use of Reclaimed Water, Article III. Reclaimed Water Distribution System. https://library.municode.com/index.aspx?clientId=16289.

Parabicoli, S. 2001. "How to Sell Reclaimed Water—Setting the Right Price in Maui." *Water 21* (August): 30.

Parabicoli, S., and C. Brown. 2004. *The Growth of Water Reuse in Hawai'i*. Proceedings of the 19th Annual Water Reuse Symposium—Bringing Life to the Desert, Phoenix, Arizona, September 19–22.

SEVEN

Catching the (Energy) Wave of the Future

LUIS VEGA AND REZA GHORBANI

Ocean waves offer a unique and promising opportunity to capture renewable energy for islands and coastal communities worldwide. In fossil-fuel-dependent Hawai'i, where 85–90 percent of energy needs are met by imported oil, more and more renewable energy options are gaining momentum. Wave energy is currently being tested and explored as a viable alternative.

Ocean waves exhibit considerable energy, which can be captured and converted to electricity to meet the needs of coastal populations. Although much wave energy exists in deep seas far away from the coast, the high cost of construction, maintenance, transmission, and storage makes it preferable to install wave energy converters (WECs) near the shore. Moreover, despite the abundance of wave energy in the sea, technologies for its extraction need to overcome major challenges, including intermittence in wave energy supply and survivability of installations in stormy weather. Current wave energy devices use a range of methods to survive extreme conditions but are perhaps not quite robust enough (Cruz 2008). Mastering the two challenges of intermittence and survivability has been among the major objectives of research and development in recent years, with varying degrees of success, and will remain objectives for years to come. This chapter provides an overview of the main technology trends of WECs, the challenges and barriers to their use, and the resource of wave energy in Hawai'i.

WAVE ENERGY CONVERTERS: DEVICES THAT CONVERT
WAVE ENERGY TO ELECTRICITY

There are many different conceptual designs of WECs. There are floats, flaps, ramps, funnels, cylinders, airbags, and liquid pistons. Devices can be at the surface, on the seabed, or anywhere between. They can face backwards, forwards, sideways, or obliquely and move in heave, surge, sway, pitch, and roll motions. They can use oil, air, water, gearing, or electromagnetics for generation. WECs make a range of different demands on attachments to the seabed and connections of power cables.

One type of WEC uses waves to send water to an elevated reservoir. The stored water is then released through a turbine at a lower elevation. Others convert wave energy directly to the oscillatory motion of a rigid body, which then drives a linear generator. The body can be a cam, a buoy, an Archimedes Wave Swing, or a series of rafts hinged together at the ends. Alternatively, waves can cause the water surface inside a fixed chamber to rise and fall, which forces the air above it through a turbine.

Various stages (fig. 7.1) are required to convert the energy in the wave to electricity (Price 2009). The first stage transfers the wave energy to the WEC's body (for example, oscillating a float, rotating a flap, or compressing air). Most WEC designs have gearing in this stage to convert wave-induced motion, characterized by low velocities and high forces, to motion with high velocities and low forces. This increases the efficiency of electricity generation. The second stage transfers the WEC body's power to the power take-off (PTO) stage. This can be accomplished by pumping oil, turning a turbine, or moving an electromagnetic generator. The third stage transfers the power from the PTO to a useful form of energy (for example, generating electricity, creating water head pressure, or causing the flow of a liquid). Energy storage and power conditioning may take place in any or all of three stages of power transfer to convert a random or cyclical flow of power, which waves exhibit, into a smoother power output (Soloway and Haley, 1996).

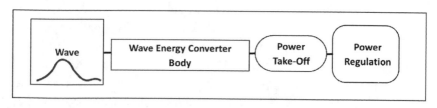

FIGURE 7.1. Power transfer in a generic WEC.

WECs currently applicable in the state of Hawai'i can be categorized under two operating principles: oscillating water column (OWC) and wave activated. An OWC device includes an air compression chamber in which a rising and falling water surface, caused by waves, produces an oscillating air current. A turbine then generates electricity using the air current. OWCs often use a low-pressure air turbine that rotates continuously in one direction regardless of the direction of the air flow. Power regulation consists of a series of power transformations, including storage. In an OWC, storage can take place only after electrical conversion, as the power is already in electrical form. Wave-activated devices work in one of three ways: (1) oscillate due to wave action relative to a fixed part of the device and use a hydraulic system to turn a motor-generator; (2) use a linear generator, which generates electricity by moving a magnetic assembly within a coil; or (3) use direct rack-and-pinion mechanical coupling. Two well-known WECs under development are the Achimedes Wave Swing and Pelamis (Price 2009). The Archimedes Wave Swing has an electromagnetic direct-drive generator that generates "wild" AC electricity, variable moment by moment in voltage and frequency. An electronic regulation component converts that wild AC electricity to DC, in a step known as rectification, and then back to AC at a frequency that can be sent to the electrical grid for distribution. If a steady power level is required, then storage can be included in the Archimedes Wave Swing design.

In the Pelamis WEC, the power conditioning function is carried out at all three stages of energy transfer. In the first stage, the energy in large waves is not translated into motion. Large waves cause submergence of the device. In the power capture stage, forces can be limited through the choice of power that is absorbed by the PTO. The PTO mechanism is a hydraulic pump that works with both low and high pressures and rectifies the power generated. To regulate the power, the high and low pressures are stored in accumulators, which feed a high-speed hydraulic rotary machine. This process does the job of storage and gearing. Excess power can be dissipated using a resistive water heater to regulate the power.

LICENSING, PERMITTING, AND OTHER CHALLENGES TO IMPLEMENTING WAVE ENERGY DEVICES

The proposed location of a WEC device determines which regulations, and which agencies, must approve the site and the device. At least six federal agencies are involved, including the Federal Energy Regulatory Commission

(FERC), the Bureau of Ocean Energy Management (BOEM),[1] the Army Corps of Engineers, the US Environmental Protection Agency (EPA), the National Oceanic and Atmospheric Administration (NOAA),[2] and the US Coast Guard. In addition, various state, county, and city agencies may also be involved. The project must comply with several other applicable laws beyond the licenses and permits that must be secured from these different agencies.

Independent of location, licensing of wave energy devices is the responsibility of FERC. In Hawai'i, the state government has jurisdiction up to three nautical miles offshore. The federal government (embodied in BOEM) has jurisdiction in the "outer continental shelf," which includes the submerged lands, subsoil, and seabed between the outer limits of state waters and the inner boundary of international waters, which begins approximately 200 nautical miles offshore.

For wave energy projects to be located on the outer continental shelf, BOEM will issue leases, easements, and rights-of-way and will conduct any necessary environmental reviews, including those under the National Environmental Policy Act (NEPA). FERC has exclusive jurisdiction to issue licenses and exemptions for the construction and operation of wave energy projects and will conduct any necessary analyses, including those under NEPA, related to those projects. FERC, however, will *not* issue a license or exemption until the applicant has first obtained a lease, easement, or right-of-way from BOEM. Moreover, BOEM and FERC can choose to become cooperating agencies in the preparation of any environmental document required under either process. This does not preclude other agencies within the Department of the Interior (such as the US Fish and Wildlife Service, the National Park Service, and the Bureau of Indian Affairs) from intervening. This situation could lead to the requirement of two distinct (although similar in content) environmental impact statements: one for BOEM, and another for FERC. For wave energy projects located in state waters (within three nautical miles of the coastline), BOEM has no jurisdiction; licenses would still be issued by FERC, and all other requirements would be under state, county, and city rules.

This bureaucratic and regulatory "alphabet soup" represents a significant impediment to the deployment, testing, and eventual implementation of wave energy devices not just in Hawai'i but elsewhere in the United States and the world. There is a need to streamline the burdensome, although necessary, process of obtaining licenses and permits, including the necessary environmental impact statement(s). The process is project specific and expensive and requires about three years for commercial projects. The development of a "one-stop shop," where industry can process all of the documentation for licensing and permitting under federal, state, city, and county regulations, could help meet this need and ease the bureaucratic burden. Such a centralized agency must also help industry and government clear up contradictory requirements and interdepartmental jurisdictional disputes and must avoid duplicity.

Despite these bureaucratic hurdles, some WEC systems are in the precommercial phase of development and are being deployed to test and demonstrate their feasibility. Several experimental projects have already demonstrated the ability to convert wave energy into electrical energy but lack the operational records required to proceed into commercialization. Adequately sized pilot or precommercial projects must be implemented to obtain these long-term operational records. In addition, validation of the performance and survivability of specific WECs under harsh ocean conditions is required to gain commercial acceptance.

Some WEC designs have appropriate operational records, although they are cost-competitive only under limited conditions. A validated concept is given by, for example, incorporating the OWC LIMPET WEC into new breakwaters (Islay LIMPET Wave Power Plant 2002). Currently, this concept is not practical in Hawai'i because no new breakwaters are planned. Major challenges to WEC implementation in Hawai'i can be summarized as follows:

- To our knowledge, there are no first generation WEC systems that would be cost-competitive in Hawai'i.
- There is a lack of consistent funding, which would be required for industry to proceed from concept design to the required precommercial demonstration phase.
- The aforementioned regulatory process, with its possibility for overlapping and contradictory requirements, acts as a brake on innovation and implementation.

Perhaps a lesson can be learned from the successful commercialization of wind energy in Germany, Denmark, and Spain. Those successes were due to consistent government funding of pilot or precommercial projects, which led to appropriate and realistic determination of technical requirements and operational costs. By "commercialization," we mean that equipment can be financed under terms that yield cost-competitive electricity. This, of course, depends on site-specific conditions and electricity prices.

One assessment of wave energy resources and their economics (Hagerman 1992) estimates that wave energy could be a meaningful component of Hawai'i's energy production, at least for parts of the state. The report states: "Except for O'ahu, where electricity demand is comparable to 2/3 of the resource base, wave energy can be withdrawn at very low levels and still make a *reasonable contribution* to energy needs in the State of Hawai'i" (1992, 2–38). This is a seminal report, and it remains the main reference for estimates of the wave resource, as well as the identification of the coastal segments exposed to the relatively highest resource. Unfortunately, Hagerman did not address siting issues and resource seasonal variability.

Hagerman (1992) estimated the average power present in waves along coastal segments (table 7.1) on the islands of Kaua'i, O'ahu, Moloka'i, Maui, and Hawai'i island using ocean data available as of 1991. That power is measured in kilowatts per meter and is known as wave power flux[3]. Estimated annual averages of wave power flux at eighty meters depth ranged from 10 to 15 kW/m (fig. 7.2). However, because the island shelves are so narrow, even

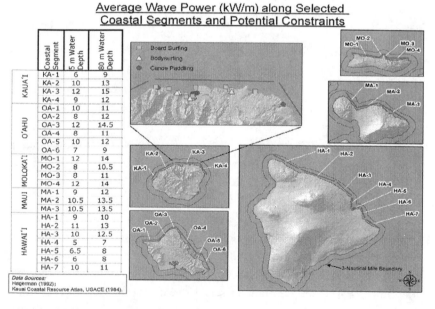

FIGURE 7.2. Wave power flux (kilowatts per meter) along coastline segments identified in Hagerman (1992).

sites with eighty-meter water depth can be closely sheltered by adjacent head-lands or peninsulas, which is the case at Kailua, Oʻahu, and in the vicinity of Hilo. At these locations, wave power density along the eighty-meter depth contour ranges from 7 to 9 kW/m. Moreover, refraction and shoaling significantly reduce wave power densities in shallower water; for example, along the fifty-meter depth contour, they were assumed to be roughly 20 percent lower than along the eighty-meter contour. Multiplying the average wave power flux along the eighty-meter depth contour by the length of each coastal segment (table 7.1), and by the number of hours in a year, and summing the results for all segments, gives an estimate of the annual wave energy resource for a particular island. Estimated wave energy resources (in gigawatt-hours per year) from the 1992 Hagerman report are given in table 7.2.

WAVE FARMS: OCEAN AREA AND SITING REQUIREMENTS

A wave farm would consist of arrays of WEC devices, spaced so that interactions between individual components are minimized. For example, about seven square kilometers of ocean area would be required for a 100-megawatt wave farm that included one hundred 1-MW WECs, or two hundred 0.5-MW WECs.[4] For comparison, a 100-megawatt offshore wind farm would require about twelve square kilometers.

To estimate the amount of electricity that could be generated from wave energy on an annual basis, a "global capacity factor" is calculated (table 7.2, second row). Given Hawaiian wave resources, and efficiencies achieved with viable WEC devices, we assume an all-encompassing global capacity factor of about 15 percent. The cumulative name plate capacity (the amount of electricity that a WEC can produce, in megawatts) and the ocean area that would be required to accommodate the arrays are given in the third row of table 7.2. It must be emphasized that this analysis ignores seasonal resource variability and assumes that all coastal segments are utilized. This is not feasible because of conflicting ocean uses and because some of these segments would be off-limits for the installation of WECs.

The potential amount of electricity that could be generated is significant. For example, the 2007 electricity demand for the islands of Hawaiʻi, Kauaʻi, and Molokaʻi could have been generated with WEC farms somehow deployed in all coastal segments. In the case of Maui, the analysis indicates that as much as 90 percent of electricity demand could be met in this fashion, but for Oʻahu, less than 17 percent. These estimates assume an annual basis, without matching

TABLE 7.1. Coastal segments exposed to predominant wave climates

Coastal Segment	Description	Length (km)
KAUAI-1	Nahili Pt. to Makaha Pt.	12
KAUAI-2	Makaha Pt. to Haena	18
KAUAI-3	Haena to Kepuhi Pt.	23
KAUAI-4	Kepuhi Pt. to Kahala Pt.	12
	Kaua'i subtotal	65
OAHU-1	Kaena Pt. to Kaiaka Pt.	18
OAHU-2	Kaiaka Pt. to 4 km SW of Kahuku Pt.	17
OAHU-3	Kahuku Pt. vicinity	8
OAHU-4	4 km SE of Kahuku Pt. to Pyramid Rock	34
OAHU-5	Pyramid Rock to Moku Manu Island	4
OAHU-6	Moku Manu Island to Makapuu Pt.	20
	O'ahu subtotal	101
MOLOKAI-1	Ilio Pt. to 6 km SW of Kahiu Pt.	26
MOLOKAI-2	West coast of Kalaupapa Peninsula	6
MOLOKAI-3	East coast of Kalaupapa Peninsula	6
MOLOKAI-4	6 km SE of Kahiu Pt. to Lamaloa Head	20
	Moloka'i subtotal	58
MAUI-1	Nakalele Pt. to Kahului	18
MAUI-2	Kahului to Opana Pt.	20
MAUI-3	Opana Pt. to Pukaulua Pt.	34
	Maui subtotal	72
HAWAII-1	Upolu Pt. to Kukuihaele	33
HAWAII-2	Kukuihaele to Laupahoehoe Pt.	36
HAWAII-3	Laupahoehoe Pt. to Pepeekeo Pt.	23
HAWAII-4	Pepeekeo Pt. to Hilo Bay	12
HAWAII-5	Hilo Bay to Leleiwi Pt.	8
HAWAII-6	Leleiwi Pt. to 3 km NW of Kaloli Pt.	12
HAWAII-7	3 km NW of Kaloli Pt. to Cape Kumukahi	16
	Hawai'i subtotal	140
	State total	436

Source: Hagerman (1992).
Note: Segments are shown in fig. 7.2.

the resource to the demand, and they assume that all electricity generated can be used when it is produced, or somehow stored for later use. These assumptions are not realistic; as shown in later sections of this chapter, the resource seasonal variability is such that during winter months electricity generation could be as much as six to seven times more than in the summer months.

Given the limited availability of unpopulated coastlines, siting of WEC devices would be challenging. In addition, WECs are currently designed to operate in waters shallower than about seventy meters, and because of the

TABLE 7.2. Theoretical wave power contribution for the generation of electricity

	Hawai'i	Maui	O'ahu	Kaua'i	Moloka'i	Total
Wave energy resource (GWh/year)	12,900	8,200	9,600	7,200	6,800	
Extractable energy with a wave-to-electricity converter (GWh/year) cf:[1] 15%	1,940	1,230	1,440	1,080	1,020	6,710
Required wave farm name plate[2]/wave farm ocean area requirement (MW/km²)	1,476/103	936/66	1,096/77	822/58	776/54	
2007 electricity consumption (GWh/year)	1,259	1,385	8,293	266	39	11,242
Potential wave energy contribution to electricity consumption	150%	90%	17%	400%	2,600%	60%

Notes: [1] Capacity factor, or the number of hours per year, expressed as a percent of 8,760 hours, during which a WEC array operates at the rated power capacity (name plate; see note 2, below). We assume that the WECs in these arrays would operate at their rated capacity 15 percent of the time; during the other 85 percent of the time, they will operate at lower-than-name-plate levels, if at all.
[2] Name plate capacity is the amount of electricity that a WEC (or, in this case, an array of WECs) can potentially generate in a year.

relatively narrow insular shelf surrounding the islands, wave farms would have to be deployed within one to three kilometers from the shoreline, in full public view. In summary, the issues of resource seasonal variability, siting considerations, and the corresponding nearshore ocean area requirements pose a daunting challenge to the implementation of wave farms in the state of Hawai'i.

OFFSHORE WAVE POWER RESOURCES IN HAWAI'I

Wave power resources off the state of Hawai'i consist of three main climate patterns: north swell, south swell, and wind waves. The Hawaiian Islands are exposed to swells from distant storms, as well as seas generated by trade winds. The island chain creates a localized weather system that modifies the wave energy resources from these distant sources. The group led by Kwok-Fai Cheung, working for the Hawai'i National Marine Renewable Energy Center of the University of Hawai'i (see http://hinmrec.hnei.hawaii.edu/), estimated the effect of local weather on wave energy resources for the Hawaiian Islands using the global WaveWatch3 model (SWAN Team 2006) and the Weather Research Forecast model. The winds and deep-water ocean waves estimated by the models compare favorably with satellite and buoy measurements, providing a measure of validity to the model predictions.

The validated model reveals that in deep water (>150 m depths), during the winter months, northwest swells have relatively large amounts of wave power, up to 60 kW/m (power per wave-crest unit length). However, in the summer months, the wave power density from northwest swells is less than 10 percent of the winter values. South swells, prevalent in the summer months, have lower power levels of less than 15 kW/m. The wind waves are the most consistent throughout the year and yield offshore power levels in the range of 5–25 kW/m. Significant seasonal variations are present at all island sites between winter and summer months.

The consistency of the wave climate and the proximity to shore play an important role in the selection of optimal locations for deployment of wave energy devices. While the north- and south-facing shores would capture swell energy, the most favorable sites are in areas exposed to the direction of the wind waves, which typically come from the northeast. Moreover, this deep-water model is not applicable to shallow-water conditions (e.g., water depths <100 m), and the WEC devices under development are to be installed in depths of no more than seventy meters. We evaluated the wave energy resource for shallow-water conditions in coastal segments selected by Hagerman (1992), which are all northeastern exposures (except one, at South Point on Hawai'i island). We estimated wave power flux at those sites using a shallow-water model (fig. 7.3), which supports Hagerman's (1992) selections and the use of north-easterly wind wave resources.

SHALLOW-WATER WAVE POWER RESOURCES IN HAWAI'I

To estimate the shallow-water resource, the SWAN wave-simulation model (SWAN Team 2006) was used with spectral wave data hindcasted from Wave-Watch3 to obtain ten years (January 2000 through December 2009) of parametric wave data required by designers of WEC devices. This project considered five representative sites: two water depths at North Beach in Kane'ohe, Pauwela in Maui (the Oceanlinx site), a shallow-water site off South Point (Hawai'i island), and a shallow-water site in the 'Alenuihaha Channel off Upolu Airport (Hawai'i island). The model was evaluated using archival data available from the stations listed below. For reference, the water depth and latitude and longitude coordinates of each site are presented in table 7.3. We

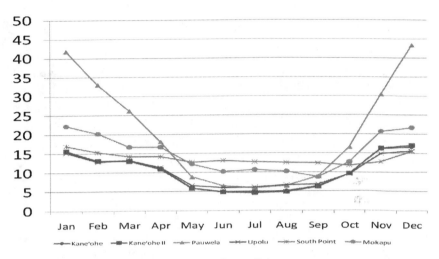

Wave Power Flux (kW/m) Monthly Average
(10-year hindcasts and 5-years Mokapu Buoy)

FIGURE 7.3. Monthly average of wave power flux at all sites.

TABLE 7.3. Location of sites selected for SWAN analysis

Station	Location	Lat. (N)	Long. (W)	Water Depth (m)
51201	Waimea Bay	21.673	158.116	198
51202	Mokapu Point	21.417	157.668	100
Kane'ohe	Kane'ohe, O'ahu	21.465	157.752	27
Kane'ohe II	Kane'ohe, O'ahu	21.472	157.747	58
Pauwela	Pauwela, Maui	20.958	156.322	73
Upolu	Upolu, Hawai'i	20.275	155.863	47
South Point	South Point, Hawai'i	18.910	155.681	40

selected the latitude and longitude, and the bathymetry is from LiDAR data with a resolution of about four meters.

These five locations were chosen to be in water depths of less than about seventy meters, to coincide with the upper limit of WEC devices currently under design by reputable developers. The Kane'ohe site coincides with the location of a 40-kilowatt WEC device ("heaving buoy"), currently installed and undergoing testing. The Kane'ohe II site was selected as a possible location for an additional WEC in deeper waters (58 m vs. 27 m). The Pauwela site (73 m depth) is within the area specified in the preliminary FERC permit awarded

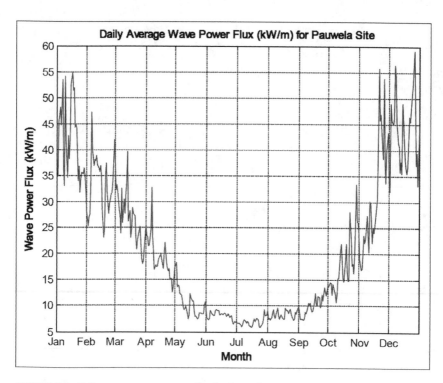

FIGURE 7.4. Daily average wave power flux for Pauwela site (73 m).

to Oceanlinx, which is currently completing the design of an oscillating water-column system that will be able to generate approximately 2.7 megawatts. The site selected by Oceanlinx in Pauwela represents a relatively high wave power resource (figs. 7.4 and 7.5).

The graphical representation in figure 7.3 is indicative of the relatively high seasonal resource variability with summer months showing power levels of one-seventh the winter values in Pauwela and one-third of that in Kaneʻohe. The other two sites off Hawaiʻi island were deemed to be interesting because of their exposure to trade wind waves (Upolu at 47 m) and both trade wind waves and southern swell (South Point at 40 m). At sites exposed to southern swell, like South Point (and Mokapu), the seasonal difference is less pronounced.

Daily and monthly variations in available wave energy (table 7.4) are also significant, and these variations need to be considered when choosing sites for WECs. Significant seasonal variations between winter and summer months are evident in the data; however, daily variability is also pronounced, and relatively

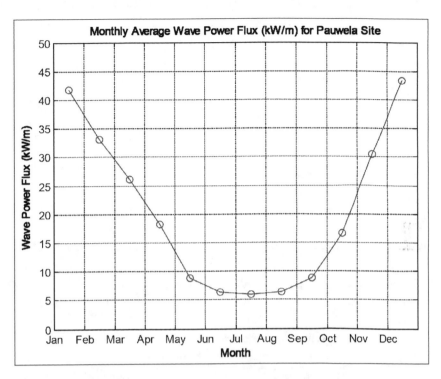

FIGURE 7.5. Monthly average wave power flux for Pauwela site (73 m).

TABLE 7.4. Monthly average wave power flux

| Site | Power Flux (kW/m) | | | | | | | | | | | |
	Jan.	Feb.	Mar.	Apr.	May	Jun.	Jul.	Aug.	Sep.	Oct.	Nov.	Dec.
Kaneʻohe	15.5	13.1	13.2	11.1	5.9	4.9	4.8	5.0	6.3	9.7	16.2	16.6
Kaneʻohe II	15.7	13.1	13.0	10.9	5.8	4.8	4.6	4.8	6.2	9.7	16.3	17.1
Pauwela	41.8	33.0	26.1	18.1	8.8	6.4	5.9	6.4	8.9	16.6	30.4	43.3
Upolu	15.2	12.8	13.2	11.3	6.5	6.0	6.1	6.6	6.8	9.6	14.9	15.6
South Point	17.0	15.3	14.2	14.2	12.7	13.1	12.7	12.5	12.4	11.8	12.8	15.5
Mokapu Buoy	22.2	20.2	16.7	16.7	12.2	10.2	10.7	10.2	8.7	12.7	20.7	21.7

large swings would be expected in the power output from any WEC device. Comparing the annual average estimates that we obtained (table 7.5) with Hagerman's 1992 estimates (fig. 7.2) indicates that, with the exception of the Pauwela site, the original estimates are useful for initial evaluation. However, monthly and daily estimates are required to proceed beyond a simple site evaluation.

TABLE 7.5. Annual average wave power flux

Site	Power Flux (kW/m)
Kaneʻohe (27 m)	10.2
Kaneʻohe II (58 m)	10.2
Pauwela (73 m)	20.5
Upolu (47 m)	10.4
South Point (40 m)	13.7
Mokapu Buoy (100 m)	15.2

SUMMARY

Wave energy is a promising and perhaps major source of potential electricity generation in Hawaiʻi. Although the resource varies by site and season and exhibits daily variability, wave energy devices of various designs, if properly planned and sited, could be reliable contributors to a mosaic of electricity sources for the state. However, there are challenges and barriers to developing wave energy. Technical and feasibility challenges will need to be overcome by demonstration projects, and consistent funding is needed to support industry as devices proceed from concept to precommercial demonstrations. It is also necessary to streamline the regulatory process: although regulation is necessary, obtaining the necessary licenses, and permits, and environmental impact statements from a bewildering (and sometimes overlapping) array of federal and state agencies remains a significant challenge to the emerging wave power industry. A third challenge is to negotiate the power purchase agreement between the industry and utilities in a win–win manner, as is now common with other forms of renewable electricity generation. Despite these challenges, the energy present in waves represents a useful resource and remains a largely untapped source of energy for the world's most isolated island chain, Hawaiʻi, as well as other coastal communities around the world.

NOTES

1. The BOEM was formerly the Minerals Management Service, part of the US Department of the Interior.

2. The National Oceanic and Atmospheric Administration is part of the US Department of Commerce.

3. The wave power flux is measured as power passing through a vertical plane of unit width perpendicular to the wave propagation direction. It is measured in kW/m, and it is used to represent wave energy resources.

4. Seven square kilometers of ocean area could comprise, for example, 11 km (6.8 miles) along the coastline by 0.6 km (0.4 miles) away from coastline or other equivalent rectangular area.

REFERENCES

Cruz, J., ed. 2008. *Ocean Wave Energy: Current Status and Future Perspectives*. Berlin: Springer.

Hagerman, G. 1992. "Wave Energy Resources and Economic Assessment for the State of Hawaii." Report prepared for the Hawai'i Department of Business, Economic Development and Tourism.

Islay LIMPET Wave Power Plant. 2002. Belfast, Ireland: Queen's University of Belfast. http://cordis.europa.eu/documents/documentlibrary/66628981EN6.pdf.

Price, A. A. E. 2009. *New Perspectives on Wave Energy Converter Control*. Edinburgh: University of Edinburgh.

Soloway, D., and P. J. Haley. 1996. "Neural Generalized Predictive Control." In *Proceedings of the 1996 IEEE International Symposium on Intelligent Control (Patras)*, 277–281.

SWAN Team. 2006. *SWAN User Manual: SWAN Cycle III Version 40.51*. Delft, Netherlands: Delft University of Technology.

EIGHT

Green Building
Integrating the Past with the Future

MATTHEW GOYKE AND JOHN BENDON

Only a few centuries ago, housing for traditional cultures around the globe had a fairly low environmental impact or, in today's terms, was "green." This "greenness" of houses was born out of necessity rather than choice. Local materials were used because that was what was available. Placement and construction of houses were tailored to local environmental conditions that worked with the wind, sun, and topography of the area. Houses were generally small, and communities were compact and walkable because of the lack of cars (Hawaii-History.org, n.d.; HawaiiAlive.org, n.d.; National Park Service 2001).

The traits that we associate with current green building practices were commonplace elements in Hawaiian housing until fairly recently. Even through the plantation age in Hawai'i, houses could be considered fairly green by today's standard. Single-wall plantation houses used few materials; natural ventilation and raised post-and-pier construction kept them cool. The homes were in tightly knit communities built around a central mill where stores, postal services, and recreational activities were all located within walking distance. Few mechanical systems meant small electrical loads and lower fuel use.

The shift away from these traditional concepts took place as electricity, technology, and modern conveniences began to enter building practices. No longer did we have to build with trade winds in mind, because we could cool buildings with air conditioning. Cars made it easy to move between communities, and development spread to untouched land. Local materials were no longer needed as newer and cheaper building materials could be imported. Irrigation allowed for green lawns and beautiful plants, no matter what type of plants they were. The modern house and office were born and could be built almost anywhere, regardless of climate or local conditions.

As technology and modern conveniences permeated society, the impact of buildings on the environment and on human health also grew. More raw land was developed, more fossil fuels were required to operate and maintain buildings, and new problems like "sick building syndrome" began to occur as people spent more time inside buildings that had poor indoor air quality. All of these issues were items that helped to trigger the modern movement to green building rating systems.

Globally, the impact of the built environment today is staggering. Impacts include energy usage, water consumption, and waste generation. For example, estimates range from four billion to five billion gallons of potable water used each day to flush toilets in the United States alone (Trinity River Audubon Center, n.d.; National Geographic, n.d.)—almost two trillion gallons of potable water annually. If one in every ten American homes upgraded a full bathroom with WaterSense labeled fixtures,[1] almost seventy-four billion gallons of water, and about $1.6 billion, could be saved per year (US Environmental Protection Agency 2010). Meanwhile, typical North American commercial construction practices can generate 2.5 pounds of waste per square foot of floor space built (USGBCWV, n.d.). This results in 250,000 pounds of solid waste generated for a 100,000-square-foot project.

Hawai'i is the most fossil-fuel-dependent state in the nation. Like many island states and nations, Hawai'i is particularly dependent on fossil fuels for transportation and the majority of its energy production. Hawai'i residents experience the direct economic impact of fossil-fuel prices at the pump and the plug: electricity costs per kilowatt hour in Hawai'i are typically three to four times higher than in the rest of the United States. It is clear that Hawai'i must reduce its dependence on fossil fuel. On January 31, 2008, Hawai'i governor Linda Lingle signed a memorandum of understanding with the US Department of Energy for the Hawai'i–DOE Clean Energy Initiative (Hawai'i Clean Energy Initiative 2014). The goals of this initiative are to reduce energy demand and to accelerate the use of renewable energy resources in Hawai'i in residential, building, industrial, utility, and transportation end-use sectors, such that a combination of efficiency measures and renewable energy generation will be sufficient to meet 70 percent of Hawai'i's energy demand by 2030. This is a bold yet achievable undertaking. One component to achieving this goal is sustainable building design and construction.

Sustainable best practices can significantly reduce the impacts of the built environment on energy consumption, water usage, and waste generation. Sustainable practices can be adopted for design, construction, and operations—in other words, the entire life of a building—and provide benefits to developers, owners, and operators. Such benefits include reduced operating costs from reduced energy and water consumption, enhanced marketability, increased worker productivity, and possibly improved employee recruitment and retention.

This chapter uses a local residential case study to showcase sustainable building practices of a real project in Hawai'i. The Kumuhau Development is a project of the Department of Hawaiian Home Lands that is located in Waimanalo on O'ahu. This project received the *Green Builder Magazine* Home of the Year award for 2011. The Kumuhau Development comprised thirty-nine entry-level homes ranging from three to five bedrooms. The team included the following key sustainability partners: architect, builder/contractor, green architectural consultant, sustainability consultant, and green rater.

THE DESIGN PROCESS

The most successful sustainable projects begin by integrating sustainability principles from the initial concept of the project; success is measured by the quantity of sustainable measures that were incorporated into the project in a cost-effective manner to keep the project viable. A system known as Integrated Design Process (IDP) focuses on very early team participation and full integration into the project, looking for synergies, efficiencies, and better ways to achieve project goals.

The standard design process is very linear:

1. The architect designs the look and feel of the building.
2. Engineers add the electrical and mechanical systems within the spaces designated by the architect.
3. Contractors build what was designed by the architects and engineers without contractor input.
4. Operators inherit, operate, and maintain the building (they often have little or no input into the design of the building).

IDP, by contrast, has all the key players at the beginning of the project. The owners, operators, engineers, architects, contractors, and sustainability consultants are all at the discussion table from the start, incorporating strategic building best practices into the very concept of the project. They help determine which systems to use and the best placements of those systems. This different approach to the design process can optimize plumbing systems, thus eliminating wasted piping, strategically place the HVAC system to minimize the number and length of ducts, and minimize waste generation by planning for the efficient use of materials.

IDP is a challenge for some project teams. They must have a willingness to do things better, and different ideas must be embraced. Sometimes, obvious solutions do not come from the expert in a particular area but are pointed out by a fresh pair of eyes that have not been implementing the same systems and procedures for long periods of time. The composition of the project team also has a significant impact on the ultimate success of a project. IDP is a positive, integrated approach with all team members meeting on common ground to develop project goals together.

The American Institute of Architects has developed a series of agreements that delineate the roles and responsibilities associated with IDP. In addition to broader-based design-phase collaboration, other changes from traditional design-bid-build project delivery methods include aligning the participants' interests to share in the risks and rewards of the project, communicating openly, and budgeting and scheduling in a collaborative fashion.

ROLE OF GREEN BUILDING RATING SYSTEMS

The movement towards using green building rating systems is a relatively new phenomenon that has many roots in age-old building concepts. The core concept behind a green building rating system is that it provides the means to help design, construct, and then rate a building that is more environmentally responsible based on a set of defined metrics. Although there are a vast number of rating systems internationally, they generally address the same key areas of building impacts and performance: energy use, water use, material use, carbon emissions, location impacts, site impacts, environmental impact, indoor environmental quality, pollution, building durability, and building operation.

The first major rating system to come onto the market was BREEAM (Building Research Establishment Environmental Assessment Method), developed in the United Kingdom in 1990. The BREEAM system has rated over 200,000 buildings since it was first launched, and it remains the main rating system used in the United Kingdom. Some of the other main rating systems around the world include the following:

- CASBEE (Comprehensive Assessment System for Building Environmental Efficiency) developed in Japan, beginning in 2001
- iiSBE (International Initiative for a Sustainable Built Environment), which has the SBTool that provides a framework for local green building programs and has been used internationally
- Green Globes, a rating system developed under the GBI (Green Building Initiative) and the most widely used in Canada and the United States
- LEED® (Leadership in Energy and Environmental Design) created by the US Green Building Council (USGBC®), the most commonly used green building rating system in the United States, with increasing international exposure

LEED, created by USGBC, is the most commonly used green building rating system in the United States, and has increasing international exposure. LEED is a points-based system that has a series of prerequisites and credits that cover key target areas in efficiency and sustainability (table 8.1). There are a number of different LEED rating systems that cover different types of buildings as well as both new construction and existing buildings. The rating systems cover homes, neighborhoods, commercial buildings, and interiors; existing commercial buildings' operations and maintenance; and more. The requirements of the rating systems are outlined in a series of reference guides. An example of the rating systems and reference guides is outlined below.

Each rating system has prerequisites that must be met for every building, as well as credits that can be completed to earn additional points. In LEED, as a project complies with criteria and earns points, those points are accumulated and lead up to a final point total. A project can achieve a designation of LEED Certified, Silver, Gold, or Platinum levels. Platinum projects have earned the most points. Each of the LEED credits and prerequisites was developed through a consensus-based process; the rating system was created with input from a variety of stakeholders across many industries, and each rating system is open for periods of public comment and review.

LEED uses a series of metrics to evaluate both the environmental impact and human benefit in core areas: sustainable sites, water efficiency, energy and atmosphere, indoor environmental quality, material use, location and linkages,

TABLE 8.1. The LEED rating system

Rating System	Reference Guide	Adaptation (full name)
LEED for Building Design and Construction	LEED Reference Guide for Building Design and Construction	LEED for Building Design and Construction: New Construction and Major Renovations
		LEED for Building Design and Construction: Core and Shell Development
		LEED for Building Design and Construction: Schools
		LEED for Building Design and Construction: Retail
		LEED for Building Design and Construction: Healthcare
		LEED for Building Design and Construction: Data Centers
		LEED for Building Design and Construction: Hospitality
		LEED for Building Design and Construction: Warehouses and Distribution Centers
	LEED Reference Guide for Homes Design and Construction	LEED for Building Design and Construction: Homes and Multifamily Lowrise
		LEED for Building Design and Construction: Multifamily Midrise
LEED for Interior Design and Construction	LEED Reference Guide for Interior Design and Construction	LEED for Interior Design and Construction: Commercial Interiors
		LEED for Interior Design and Construction: Retail
		LEED for Interior Design and Construction: Hospitality
LEED for Building Operations and Maintenance	LEED Reference Guide for Building Operations and Maintenance	LEED for Building Operations and Maintenance: Existing Buildings
		LEED for Building Operations and Maintenance: Data Centers
		LEED for Building Operations and Maintenance: Warehouses and Distribution Centers
		LEED for Building Operations and Maintenance: Hospitality
		LEED for Building Operations and Maintenance: Schools
		LEED for Building Operations and Maintenance: Retail
LEED for Neighborhood Development	LEED Reference Guide for Neighborhood Development	LEED for Neighborhood Development: Plan
		LEED for Neighborhood Development: Built Project

Source: Information courtesy of US Green Building Council.

awareness and education, innovation in design and operations, and regional priority. Many of the performance metrics are derived from certifications or standards from other organizations that specialize in specific areas. For example, energy and ventilation requirements are tied to ASHRAE (American Society of Heating, Refrigerating and Air-Conditioning Engineers) and Energy Star. Material evaluation is tied to third-party material-certifying bodies, such as Green Seal and Forest Stewardship Council. In doing this, the LEED rating system helps pull together a robust set of requirements and standards from various other organizations that are typically far beyond what current code requires for building performance in most jurisdictions.

Even though LEED is designed to always be ahead of building codes, the rise in popularity of voluntary green building rating system standards has also helped to spark an improvement in green codes and standards. In 2010, the International Code Council rolled out the IGCC (International Green Construction Code), and earlier that year, ASHRAE released a tougher green building standard (189.1). The codes and standards are meant to address the core prerequisite requirements outlined in green building rating systems like LEED and are meant to fit into the current code structures in different jurisdictions and building departments. By doing this, building inspectors are now more involved in helping to check and inspect core metrics (the importance of inspections and testing is identified in the next section).

The main rating system focus in this chapter is the USGBC LEED certification program.[2] The Kumuhau Development described below used the LEED for Homes rating system to certify the project (US Green Building Council 2009). As with most green building rating systems, LEED for Homes uses a series of points and prerequisites in different credit category areas to determine a final certification level. One key difference in LEED for Homes is that the certification level is based on an adjusted threshold, which is determined by the size and number of bedrooms in a house: the system rewards smaller homes.[3] Smaller homes are beneficial in island locations like Hawai'i because they generally use fewer finite resources, such as energy and water. They also use less material, which can mean less material shipped and ultimately ending up in landfills.

THE KUMUHAU COMMUNITY

General sustainability goals were established by the Department of Hawaiian Home Lands in their request for proposals for the Kumuhau community. The stated goal was a 3-Star Rating using the Hawaiʻi BuiltGreen rating system, which at the time was the local self-certified system.[4] The builders wanted to exceed the minimum goals and decided to evaluate LEED for Homes for the Kumuhau project. A green architectural consultant was added to the team to guide the sustainability process. Design charettes, or architectural planning sessions, were held over multiple days to determine the goals of the project. Key team members at the design charettes were architects, consulting sustainable architects, a LEED for Homes accredited professional, the developer, and the contractor. At the end of the design charette, the team had made the following key decisions:

- LEED for Homes was a good fit for the project.
- A LEED-H Silver Certification level would be the project commitment.
- Homes would be designed primarily for natural ventilation. Whole-house fans would be added to each home for mechanically assisted ventilation when needed.
- Homes would be designed for abundant natural daylighting.
- Homes would be designed for local lifestyle with an indoor/outdoor design.
- Homes would be designed to allow for community and social interaction and would not act as a barrier to human interactions.
- Carports would be standard for each home, with a solar water heating tank and washer and dryer accommodations located in the carport.
- Solar water heaters, photovoltaic systems, and rainwater harvesting would be used.
- Solar clothes dryers would be provided, as well as accommodation for electric clothes dryers.

Sustainable Design Is Place Appropriate

Some developments in Hawaiʻi are incorporating elements of sustainability by adding on such items as solar hot water, photovoltaic panels or photovoltaic-ready areas, or higher-efficiency insulation, such as expanding foam. While these strategies can be a good thing and contribute to a project's sustainability, they can also be considered "greenwashing": making projects seem more sustainable than they actually are. Taking standard projects and tacking on specific strategies can raise the bar slightly but will never reap the benefits of incorporating sustainability from the project inception and determining the most cost-effective measures for the project. First and foremost, a project should be designed to be place appropriate. The design needs to encompass the climate and microclimate of the area to the fullest degree possible. An inventory of resources should be conducted to determine what natural resources can be harvested and utilized for each project.

The Kumuhau project in Waimanalo is located on the eastern side of Oʻahu. Waimanalo has a large, flat valley that extends from the white sandy shore on the eastern side to the base of the mountains on the western side. The climate in Waimanalo consists of hot, humid temperatures, with many passing showers carried by trade winds. Several passive design strategies benefit homes located in this type of climate.

The site layout at Kumuhau had already been established, so the design team was not able to optimize the orientation of the homes. The design team could only respond to the environmental conditions through manipulation of the structure. The most important decision was to eliminate the use of air-conditioning. All subsequent decisions were based on creating comfortable living environments through natural ventilation.

The staggering of homes during the site planning phase is a strategy that maximizes the access to wind for as many homes as possible. Although the site plan was established by others, the design team strived to place two-story homes next to one-story homes, so one home did not block wind to the next home. The site planning allowed for homes to be staggered across the street, even though the depth of the lots did not allow for homes to be staggered effectively along the length of the street.

STRATEGY 1: CREATING OUTDOOR USABLE SPACE: The climate in Waimanalo is ideal for outdoor living, so the design team decided to incorporate usable outdoor spaces, such as with carports. Using a carport in lieu of a garage accomplishes many sustainable goals. By placing the homes farther away from the street, cars could park in the driveways, which opened up the carports for other activities.

STRATEGY 2: WINDOW SIZE, QUANTITY, AND PLACEMENT: The design team placed large openings in every room except the bathrooms. Windows were placed on adjacent walls so wind could flow through the room. When possible, awnings were used; when awnings were not possible, windows were protected by an eave so that they could remain open during frequent passing showers. Windows or doors without roof overhang protection were provided with fixed louvered sunshades. The windows themselves were dual-glazed low-emissivity (low-e) coated. Screens were provided for all openings. In addition to increased ventilation, this window and fixed sunshade strategy allowed light to come in and fill the spaces while minimizing heat entering into the home.

STRATEGY 3: REFLECTIVE ROOFING: Comfortable living environments keep heat out, increase ventilation and airflow, allow light in, and control humidity. High-albedo, or highly reflective, roof materials help to keep heat out by minimizing the absorption of heat.

STRATEGY 4: ENVELOPE—INSULATION: Insulation reduces the amount of heat that enters the home. In Kumuhau, the roof ceiling assembly and walls were insulated with R30 batt insulation. Proper installation of insulation is critical, and this home met a grade I installation, which meant there were minimal voids or gaps in the insulation, and it was not compressed to a significant degree. The insulation was inspected prior to the installation of the drywall.

STRATEGY 5: ENVELOPE—INFILTRATION: Heat can also be transferred through air movement or infiltration: air moves through cracks and gaps in the exterior envelope of a home into the walls and into the home through outlets, switches, and other gaps. A building wrap was used on this project to help reduce air infiltration. In addition, gaps around pipes, conduits, and ducts were sealed with expanding foam. This was done on the exterior envelope, interior walls, and interior floors. Although infiltration is far more important in air-conditioned homes than in naturally ventilated homes, a durability component is associated with air infiltration: air carries moisture, as well as heat, so sealing all the joints and gaps also minimizes the amount of moisture that can get into the interstitial spaces of the home.

STRATEGY 6: MECHANICALLY ASSISTED VENTILATION: In Hawai'i, the trade winds tend to slow down in the evenings, and at times they may stop completely. At these times, natural forces need assistance, which came in several forms for this project: ceiling fans, a whole-house fan, and exhaust fans. Ceiling fans were used to circulate air in individual rooms. In terms of human comfort, moving air can be at a higher temperature than stagnant air and still feel comfortable: moving air evaporates moisture from the skin, making people "feel" cooler.

Whole-house fans work on a larger scale: they pull air in from the outside through all of the windows and push it out through the attic. They can exchange the entire volume of the home in several minutes. Several things are happening while the whole-house fan is on. Cooler air is brought in from the lower parts of the home, and hotter air is pushed out through the roof. This enhances the "stack effect": the natural flow of air and heat in that same direction. The whole-house fan also replaces the air in the attic space—which can reach 140°F or more—with cooler air from the home, helping to keep the whole home cooler. Again, as with ceiling fans, air movement makes inhabitants "feel" cooler inside the home.

The Kumuhau homes were equipped with efficient, Energy Star–rated exhaust fans in the kitchens and bathrooms regardless of, or in addition to, operable windows.[5] Each fan was ducted directly to the exterior and helped to remove moisture and heat-laden air from the home. Humidity levels affect human comfort: most people feel comfortable between 40 percent and 60 percent relative humidity. In Hawai'i, the outdoor air is typically between 50 percent to 60 percent relative humidity, even on a sunny day, so the interior air of a naturally ventilated home will be about the same as the outside air. Therefore, building occupants want to avoid adding more moisture and heat into the air whenever possible. Kumuhau homes were equipped with exhaust fans in the bathrooms, which were tied to the light switches and had timers to keep the fans on for several minutes after the bathrooms were used. When showers were in use, the fans pulled the moist air out of the rooms, and after showering, the fans continued removing warm moist air for several minutes.[6] All of these strategies worked in combination to create a comfortable living environment without air conditioning, which saves a tremendous amount of energy.

SPECIFIC ENERGY EFFICIENCY STRATEGIES: The best way to approach optimizing energy efficiency in the built environment is to avoid trying to make basic building designs better and, instead, to ask four simple questions:

1. What can be eliminated?
2. What can be reduced?
3. What is the optimal placement of everything for efficiency?
4. What are the most energy-efficient versions of the items needed?

People tend to forget that the simple act of not turning on a light saves energy. Hanging laundry outside to dry naturally saves energy by not using a clothes dryer. Hanging laundry outside also has the added benefits of avoiding additional heat and humidity inside the house that will then need to be managed and mitigated. Key questions asked in the Kumuhau project included the following:

1. Is a clothes dryer necessary? Most families will want or expect a clothes dryer, so accommodations were made to allow the family to install a clothes dryer if desired.
2. Could the use of the clothes dryer be reduced? Yes, if an alternate solution was provided and convenient. Therefore, clotheslines were provided and conveniently located near the clothes washer.
3. What was the optimal placement of the washer, dryer, and clothes line? The project team wanted to keep the heat and humidity from the clothes washer and dryer out of the conditioned space. The clothes washer and dryer needed to be in a protected area, convenient to use, and not too far to carry laundry. The carport was selected as the ideal location. The appliances were placed inside a protective cabinet for aesthetic reasons and to protect them from the elements. The clothes line was placed outside in the yard close to the home.
4. What was the most energy-efficient version of the appliances? Appliances needed to be selected based on water use per load and energy efficiency. It was decided that appliances should carry the Energy Star label for energy efficiency and use as few gallons of water per load as possible.

Some of the answers to the four questions on the Kumuhau project are as follows:

1. What can be eliminated? Air conditioning.
2. What can be reduced? The number of lights needed during the day.
3. What is the optimal placement of everything for efficiency? Clothes washer outside, and sunscreens over south-facing windows.
4. What are the most energy-efficient versions of the items needed? Energy Star exhaust fans and solar water heaters.

Role of Building Commissioning Agents, HERS Raters, and Third-Party Inspections and Testing

One of the requirements in a LEED certification is building commissioning of the fundamental energy-using systems. In many island locations, including

Hawai'i, building commissioning is a new concept. As defined by the Building Commissioning Association (n.d.), "Building commissioning provides documented confirmation that building systems function according to criteria set forth in the project documents to satisfy the owner's operational needs."

This typically means that a third party inspects and does performance testing of various systems ("commissions" them) to make sure that they are operating properly. Examples of these tests might include measuring airflows from air conditioning supply registers to make sure they are meeting the design criteria, or checking lighting controls to make sure they are working properly. One of the reasons that building commissioning is a relatively new concept is that it is an additional cost to the project, and building owners expect their design and construction teams to build buildings that operate the way they are intended to. Problems can arise, however, because of the complex nature of building construction. Buildings consist of thousands of individual components that can be put together by hundreds of different people. The mix of subcontractors from different trade groups, and the complexity of putting all the pieces together, means that mistakes can and do happen. The process of commissioning is designed to help catch and identify any potential mistakes in the core energy-using systems so that systems are working as designed. By doing so, costs associated with poor building performance can often be avoided, and the cost of commissioning can therefore provide a very good return on investment.

Commercial LEED for new construction requires that, at a minimum, the following systems are commissioned: heating, ventilation, air conditioning, and refrigeration; lighting and daylighting; domestic hot water; and renewable energy systems. The commissioning agent must also review the owner's project requirements and the basis of design. For example, an owner might require a 40 percent reduction in energy use from a set standard, and the design team might specify an efficient air conditioning system to meet that design. The commissioning agent then provides a commissioning plan for testing the air conditioning system. Once installed, the air conditioning system is tested according to that plan and may include such testing processes as testing airflows, measuring duct leakage, and checking refrigerant charge.

Although the commissioning process and performance testing add additional upfront costs for the project, the benefits typically outweigh the costs. According to a report from the Lawrence Berkeley National Laboratory (Mills 2009), building commissioning offered a number of economic and performance benefits. The report studied data on 643 commissioned commercial buildings in twenty-six states totaling 99 million square feet. Key points were the following:

- High-tech buildings are particularly cost-effective and saved large amounts of energy and emissions due to their energy-intensive natures.
- Projects employing a comprehensive approach to commissioning attained nearly twice the overall median level of savings and five times the savings of projects with a constrained approach.
- Nonenergy benefits are extensive and often offset part or all of the commissioning cost.
- Limited multiyear postcommissioning data indicate that savings often persist for a period of at least five years.
- Extrapolating the author's methods to all US nonresidential buildings yielded potential energy savings of $30 billion by the year 2030 and showed the potential to reduce annual greenhouse gas emissions by about 340 megatons of carbon dioxide each year. Delivering these benefits would generate $4 billion per year in sales and support approximately 24,000 jobs. (Mills 2009)

On the residential side, commissioning is also required for LEED, but it is done by a RESNET (Residential Energy Services Network) energy rater and follows the guidelines set forth in the Home Energy Rating System (HERS) requirements and by the Energy Star program for new homes. The residential building commissioning process typically includes the following components: a midconstruction inspection of the insulation installation, called the Thermal Bypass Checklist Inspection; testing the air conditioning ductwork at final inspection using a device called a duct blaster; testing the air leakage through the building envelope using a device called a blower door; and other miscellaneous tests, including solar hot water inspections, bathroom fan performance, ventilation airflow, refrigerant charge tests, and combustion safety testing.

The performance test results for duct and building envelope leakage are then put into a HERS energy model that the energy rater uses for the home. The energy model performs a computer simulation that compares the house built to a reference design home that meets the requirements of the 2006 International Energy Conservation Code. A score is then generated to show the energy reduction (as percent below code) for the house. A score of 100 is a home that meets code, and each number below that represents a 1 percent reduction in energy use from the code-built home. Thus, a score of 75 represents a

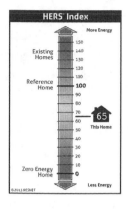

FIGURE 8.1. HERS Index (Residential Energy Services Network, n.d.).

25 percent reduction in energy use. By modeling the house, different measures can be evaluated for their effect on performance improvement.

The HERS Index (fig. 8.1; Residential Energy Services Network, n.d.), separate from the HERS energy model, is a scoring system established by RESNET in which a home built to the specifications of the HERS Reference Home (based on the 2006 International Energy Conservation Code) scores a HERS Index of 100, while a net-zero energy home scores a HERS Index of 0. The lower a home's HERS Index, the more energy efficient it is compared with the HERS Reference Home.

The Lawrence Berkeley National Laboratory has also looked at residential commissioning energy savings. Residential commissioning processes can save 15–30 percent of HVAC-related energy in existing houses and 10–20 percent in new conventional houses (Matson et al. 2002). The home inspection and commissioning process can result in benefits beyond energy savings, such as increased comfort and durability, as well as improved indoor air quality and combustion safety. As the complexity of the building systems increases, the potential benefits from building commissioning also increase. Buildings that are less complex in nature, such as those built in the Kumuhau project, will have less need for complex building commissioning procedures and thus a reduced cost for achieving LEED certification.

Because of the mild climate in Hawai'i and lack of mechanical cooling and heating in many homes, the USGBC and Energy Star programs have developed specific guidelines for the island state. Houses without air conditioning in Hawai'i are exempt from having to do many of the performance tests required for homes on the US mainland. The prerequisite requirement for the energy section of LEED homes is in compliance with Energy Star. A special Builder Option Package (Energy Star 2010) was created for Hawai'i to address the differences between Hawaiian homes and most mainland homes. Kumuhau used this Energy Star Builder Option Package approach along with some other LEED-specific inspections to meet the inspection and testing requirements for LEED.

The first major inspection for Kumuhau was a pre-drywall inspection of the insulation installation as mandated by the Thermal Bypass Checklist Inspection. The inspection process provided an opportunity for the HERS rater to train the insulation installers on the proper insulating techniques identified by the standard. Possibly because insulation is fairly new in Hawaiian homes, many of the insulation installations that were inspected in the state did not meet the required standards specified by Energy Star. Insulation that is not installed properly can lose much of its ability to insulate. In addition to inspecting the insulation, items that would later be covered up by drywall, such as fan ducts, were inspected.

The second inspection period was the final inspection, which included a check of green measures that were identified in the LEED checklist, as well as wrapping up some of the energy inspections. Mechanical systems inspected included the domestic solar hot water unit, operation of the whole-house fan, and settings and performance of bath fans. Because of the minimal mechanical systems in the house, no performance testing using a duct blaster or blower door was needed.

THE FINAL CHECKLIST

Even though minimal performance tests were needed for Kumuhau, a number of green measures still needed to be implemented by the team to achieve the LEED Gold certification. The measures implemented were nontechnical and can be done for almost any home in Hawai'i. This section reviews the final certification checklist and identifies how the Kumuhau project achieved those measures. Kumuhau earned a total of 71 points, thereby achieving a LEED Gold certification with the adjusted LEED Gold threshold of 65 points minimum.

Innovation in Design

The Innovation in Design category in LEED covers the integrated design approaches of the project and outcomes from that process, such as durability

measures and potential innovation, regional priorities, or exemplary performance in each of the credit categories. The prerequisites for all projects are that
a minimum amount of project integration occur and that a preliminary project rating be developed, with specific durability measures being listed and
checked against site-specific durability risks. The Kumuhau project team
focused on design and durability considerations early on and sought input
from a variety of project team members, including a sustainability consultant
with the LEED AP (Accredited Professional) Homes designation. Durability
measures in the project were checked by both the superintendent on the job
and the green rater on the project. Many projects in Hawai'i could benefit
from early planning that considers all elements of the house in a holistic
manner and considers specific durability measures that are then incorporated
into the plans.

Locations and Linkages

LEED recognizes that the locations of homes can play a critical role in the overall environmental impact of the home. Homes that are built in previously developed areas have less environmental impact than those built on untouched
sites known as greenfields. Sites that have amenities close by, such as parks,
restaurants, and retail areas, typically result in a reduced need for cars to get
to basic services. Proximity to public transport also helps to reduce overall car
use. It is sometimes difficult for projects in rural locations (such as many areas
in Hawai'i) to meet these criteria, which is why no prerequisites or minimum
points are required in LEED for the Location and Linkages category. However, island locations can benefit from homes meeting these criteria because
it means less land being used for new development and fewer cars on the road
using public infrastructure and consuming fuel.

Sustainable Sites

The Sustainable Sites credit category assesses the impacts of the site development, landscaping, and overall lot size. Landscaping maintenance can require
considerable resources. A year-round growing season can result in constant
watering needs and continual maintenance. According to the US Environmental Protection Agency (EPA; 2007), "Americans spend more than three
billion hours per year using lawn and garden equipment. Currently, a gasoline

powered push mower emits as much hourly pollution as 11 cars, and a riding mower emits as much as 34 cars."

Carefully planning a landscape that is regionally appropriate can help reduce water consumption. Planning with maintenance requirements in mind can help limit the impact of landscape equipment on emissions. Additionally, the landscape surrounding buildings can be designed to reduce heat island effect,[7] manage storm-water runoff from the site, and limit pests that enter the building. Plants can be used to shade hard surfaces surrounding buildings and, in turn, add to thermal comfort in and around the home. Storm water that flows off developed areas can ultimately end up in oceans, streams, or wetlands. Often storm water contains pollution that can affect fragile wetland and ocean ecosystems. Designing measures to allow water to infiltrate onsite helps to reduce the impact of developed areas. Pests like to live on plants and can climb into buildings when plants touch those buildings. Locating plants and irrigation away from buildings can help to reduce the number of pests and thus the need for chemical pest treatment.

Kumuhau implemented a number of these measures with an important focus on storm-water design. The small lots had minimal hardscape areas so that water could infiltrate onsite. A rain barrel was installed to help capture some of the water from the roof gutters so it could be used later. Small retaining walls helped to minimize erosion and runoff onto adjoining properties.

Water Efficiency

Fresh water is a precious resource in island communities. It not only is a scarce commodity but also takes a lot of energy to treat and transport to buildings. LEED for Homes addresses both indoor and outdoor water use. Simply choosing appropriate interior water fixtures can save a significant amount of water over the life of the home. According to the US EPA, each American uses an average of one hundred gallons of water a day at home. They could use 30 percent less water by installing water-efficient fixtures and appliances, such as WaterSense fixtures. Low-flow, high-efficiency fixtures are an easy alternative to standard fixtures and can be very easily incorporated into new construction or retrofit projects.

Landscape water needs can vary significantly depending on location and plant selection. In some areas, rainwater can be effectively used and captured. A few locations in Hawaiʻi, such as those close to Hilo on Hawaiʻi and some areas of Maui, use rainwater harvesting exclusively for all water uses. Other locations can take advantage of gray water, which is water that has been used once and is fairly clean (e.g., shower and lavatory faucet water) or has been treated by a municipal sewage treatment plant and is used again for irrigation needs (see chapter 6). In areas that require irrigation, natural water systems (see chapter 5) can help deliver water when and where it is needed so that it can be conserved. The Kumuhau project planted drought-resistant grass that did not require permanent irrigation because of the wet local climate. The project also used WaterSense toilets and lavatory faucets, as well as low-flow, high-efficiency showerheads.

Energy and Atmosphere

The largest credit category in the LEED rating systems is Energy and Atmosphere. Over the life of the building, the energy use of that building plays a major role in its environmental impact and the amount of greenhouse gases associated with it. Likewise, the refrigerants used to cool homes can affect both atmospheric ozone and greenhouse gas emissions. Any measures that can reduce energy consumption, coupled with energy production from renewable energy systems, constitute a large step in the right direction towards island sustainability.

The Kumuhau project incorporated a wide variety of energy-efficient measures into the house and documented predicted performance and HERS score using an energy model. A number of measures were used to mitigate heat gain and eliminate any need for heating or cooling systems. Other simple systems, such as compact fluorescent lights (CFLs), solar hot water, and Energy Star fans, helped to mitigate energy use. To offset the energy that would be used, a 2.58-kilowatt photovoltaic system (twelve panels with a rating of 215 watts each) was installed. In combination, these factors should offset the majority of the energy needs of the home.

Materials and Resources

Some of the most common elements associated with green building are the materials that are used in the building. The LEED Homes rating system looks at both the type of materials used and any ability to reduce the amount of materials used to build the structure of the house, thereby resulting in waste reduction. The program evaluates materials in three ways. Environmentally preferable products are those that are rapidly renewable (e.g., bamboo, cotton, or linoleum), have recycled content (e.g., composite decking made from recycled plastic), or are Forest Stewardship Council (FSC) certified. All tropical wood used on a project is required to be FSC certified to help curb tropical deforestation.

Another metric used to analyze products incorporates emissions specifications for those products. Many of the materials used in buildings can actually be quite toxic. Using materials that will not emit volatile organic compounds (VOCs) is considered preferable. Finally, locally produced materials, defined as materials that are sourced and manufactured within 500 miles of the project site, are incentivized. It can be difficult to source local products in island locations such as Hawai'i. Carefully evaluating options, using as few materials as possible, and reusing materials whenever possible can make a significant difference on the material impacts of buildings.

The Kumuhau project used a process of offsite fabrication of all of the wall, floor, and truss systems to help eliminate waste. By building these elements in a factory setting, minimal waste was created, and recycling options were facilitated. The project obtained points in each of the materials credit evaluation sections. An insulation that contains a minimum of 30 percent postconsumer recycled content was used throughout the house. It was also a low emission product, which helped the team to receive points in the emissions section, as did the use of low-VOC paint. In addition, the aggregate used in the concrete foundation was sourced locally.

Indoor Environmental Quality

According to the US EPA, Americans typically spend about 90 percent of their time indoors. While this estimate is probably high for those living in mild tropical environments, a significant amount of time is still spent indoors. If houses are not constructed properly, pollutants such as carbon monoxide, mold, and particulates can get trapped inside the house and make the air quality inside worse than that of the air outside. This can potentially lead to health problems over time. Often, building construction defects, or installation techniques that lead to poor indoor air quality, can also lead to comfort-related problems when mechanical systems are not operating correctly. The Indoor Environmental Quality section focuses on ways to mitigate air quality and comfort problems in the house. This mitigation is typically achieved by ensuring the mechanical systems are designed and operating properly, that the house has proper ventilation, and that any pollution buildup from construction is flushed out prior to occupancy.

For the Kumuhau project, there was no air conditioning installed, and so many of the issues related to comfort and indoor air quality concerns from poor air conditioning installation were avoided. The Kumuhau project did seek to ventilate the house with Panasonic Whisper Green bath fans that were set to run constantly at a low-flow rate. This process continually flushed the house of any pollutants that might build up. The fact that the homes only had a carport, instead of a garage, also increased the points.[8]

Awareness and Education

There are many cases in which homeowners know more about the operation of their cellular phone than they know about the major systems in their homes. The Awareness and Education section is designed to ensure that homeowners receive a robust operations manual for their home and training on how to operate the features found within the home. Knowing how often to service a solar hot water system, clean photovoltaic panels, and conduct regular maintenance on the house can result in a dramatic effect on the longevity and performance of the systems. The homeowner manual that outlines the critical systems of the home is then something that can be passed down as ownership of the house changes over time. When homeowners understand the operation and maintenance requirements of their homes, their homes will perform better over time.

Every homeowner at the Kumuhau project received a comprehensive homeowner manual that provided information regarding their home. The manual went beyond the standard compilation of installed product manuals and provided operation and maintenance procedures for the various green features of the home, such as the solar hot water, photovoltaic, and whole-house fans. Additionally, the homeowners received a walk-through training where the builder showed the occupants the various items that were included in the manual. The Kumuhau manual was intended to benefit the current occupants and also future homeowners so that proper operation and maintenance procedures continued throughout the life of the home.

CONCLUSION

Kumuhau achieved a LEED Gold certification, which exceeded the project goal of LEED Silver. This was achieved because the IDP process was implemented early, and the entire team was committed to the project. It is standard practice to include a safety factor of at least 10 percent over the points required to meet the established LEED goal, to allow for unforeseen circumstances during construction. The 10 percent buffer typically allowed the team to achieve the goals of the project even if unforeseen minor events occur. For Kumuhau, the 10 percent safety factor allowed the project to achieve LEED Gold and to exceed the expectations of the client. This was feasible only because of the excellent management of the process and to the contractors' and subcontractors' full commitment to the project and LEED goals.

Reducing the resource demands of modern homes in Hawai'i helps meet modern state requirements, but although modern homes like those at Kumuhau are more technologically advanced than native Hawaiian homes or those from the plantation era, present-day "green building" designs reflect lessons learned by the earliest Hawaiians. To be habitable, and to be sustainable, homes must fit into the local ecology, using local products and local environmental conditions to their advantage. Optimal siting allows homes to maximize natural light and ventilation, just as traditional Hawaiian homes did, and many of today's efforts to reduce resource use in the home echo the "waste not, want not" ethic that island living once demanded of the Hawaiian people. Building practices that import too many materials, generate excessive waste, import exotic landscaping materials, and offer suboptimal living conditions inside buildings—practices that have been common in Hawai'i until recently—are inherently unsustainable. It should come as no surprise that award-winning "green" buildings, such as those at Kumuhau, ultimately converge on some of the same underlying design principles that reflected centuries of experience with these islands. Using green rating systems to build more sustainable buildings combines today's technologies with strategies from the past that utilize the natural elements and results in buildings with lower environmental impacts that provide healthier, more efficient spaces for island societies.

NOTES

1. WaterSense is a program of the US Environmental Protection Agency (EPA; see http://www.epa.gov/watersense/). The WaterSense program tests and labels products and provides a convenient and easy way to find water-conserving fixtures.

2. The US government and a variety of states require LEED certification for government buildings. In Hawai'i, all state-owned or -funded buildings over 5,000 square feet and all K–12 schools are required to meet LEED Silver or equivalent certification. A large

number of US government agencies, including most branches of the military, require LEED Silver certification.

3. Similarly, more bedrooms mean more people can be accommodated in the house, meaning more resources dedicated to housing people for that house. The goal of the adjusted threshold is to move homes away from the popular trend of single families living in massive homes. This adjustment incentivizes smaller and more compact designs by requiring fewer points for them to meet the same threshold level for certification as a larger home would.

4. Hawaiʻi BuiltGreen has since been phased out and is no longer endorsed for use.

5. Energy Star is a US government program that sets energy efficiency standards for residential and commercial buildings and appliances, among other things.

6. Fans with humidistats (moisture sensors that control the fans) are also available but were not included in the Kumuhau homes.

7. A heat island effect occurs when urban areas are hotter than nearby rural areas. For more information, see the US EPA's description at http://www.epa.gov/hiri/.

8. Garages can become quite polluted: cars emit pollutants, and homeowners typically store household chemicals and other polluting products in them.

REFERENCES

Building Commissioning Association. n.d. "Commissioning." https://netforum.avectra .com/eweb/StartPage.aspx?Site=BCA&WebCode=HomePage.

Energy Star. 2010. "Builder Option Packages for Hawaii." http://www.energystar.gov/index .cfm?c=bop.pt_bop_hawaii.

HawaiʻiAlive.org. n.d. "Shelter and Kauhale System." http://www.hawaiialive.org/topics .hp?sub=Early+Hawaiian+Society&Subtopic=3.

Hawaiʻi Clean Energy Initiative. 2014. Home page. http://www.hawaiicleanenergy initiative.org/.

HawaiiHistory.org. n.d. "Architecture." http://www.hawaiihistory.org/index.cfm?fuse action=ig.page&CategoryID=280.

Matson, N., C. Wray, I. Walker, and M. Sherman. 2002. "Potential Benefits of Commissioning California Homes." http://eetdpubs.lbl.gov/publications/12/potential -benefits-commissioning-california-homes.

Mills, E. 2009. "Building Commissioning: A Golden Opportunity for Reducing Energy Costs and Greenhouse-Gas Emissions, Lawrence Berkeley National Laboratory." http://cx.lbl.gov/2009-assessment.html.

National Geographic. n.d. "Green Living." http://greenliving.nationalgeographic.com /plastic-water-bottles/.

National Park Service. 2001. "A Cultural History of Three Traditional Hawaiian Sites on the West Coast of Hawaiʻi Island." http://www.cr.nps.gov/history/online_books/kona /history1d.htm.

Residential Energy Services Network. n.d. "Home Energy Rating." http://www.resnet.us /home-energy-ratings.

Trinity River Audubon Center. n.d. "Green Building." http://trinityriver.audubon.org /green-building.

US Environmental Protection Agency. 2007. "EPA Seeks to Limit Emissions from Lawn Mowers." http://yosemite.epa.gov/opa/admpress.nsf/e87e8bc7fd0c11f1852572a00 0650c05/0cb7669b182b145d852572c0005e415a!OpenDocument.

———. 2010. "Dreaming of a Better Bathroom?" http://www.epa.gov/watersense/pubs/ bathroom.html.

USGBCWV (United States Green Building Council West Virginia). n.d. "Green Schools." http://usgbcswva.org/green-schools?device=desktop.

US Green Building Council. 2009. *LEED for Homes Reference Guide*. Washington, DC: US Green Building Council.

NINE

Shades of Green in the Tourism Sector

Sustainability Practices and Awareness in the State of Hawai'i

LINDA COX AND JOHN CUSICK

Global economic change in the first decade of the twenty-first century focused attention on the volatility that results from economic dependency on a single sector, such as tourism. Increasing concern for the environment has also resulted in a significant shift worldwide towards addressing the challenges that tourism poses to maintaining the sustainability of the host region. As a result, visitors and residents in Hawai'i and other popular tourist destinations are now encouraged, for example, to steward natural resources, to buy and serve local products, to promote energy and water conservation, and to buy carbon offset credits as a means of ensuring increased sustainability. The backdrop to these efforts is captured in a May 2011 headline in the *Honolulu Star-Advertiser*, "Tourism Outlook: Up? Down? (It Depends on Whom You Ask)" that is representative of the diverse perspectives on the tourism sector's current performance (Schaefers 2011).

The state of Hawai'i 2050 Sustainability Task Force plan suggested taking a "bold step to put the 'stake in the ground' and set the standard for Hawai'i's sustainable future" (Social Sciences Public Policy Center 2010, 33). If successful, the transition promises healthier ecosystems and human communities. However, tourism planning for the largest economic sector in the state has been poorly coordinated despite intentions initiated in the 1970s to become "the pioneer in policy planning for sustainable tourism" (Mak 2008, 56). In the last century, mass tourism development occurred on each of the main Hawaiian Islands, and with the dawning of the twenty-first century the nature of Hawai'i's tourism sector and the expectations for the sector have changed. This chapter discusses sustainability efforts within Hawai'i's tourism sector and presents information on where the state now stands.

WHAT IS SUSTAINABLE TOURISM?

Sustainable development attempts to integrate environmental, social, and economic decision making so that present needs are met without preventing future generations from meeting their own needs (United Nations 1987). The terms "sustainable tourism" and "ecotourism" receive attention in academic literature and popular media, including in Hawai'i (Fennell 2001; Cox et al. 2008; Cusick, McClure, and Cox 2009), and although attention has been paid to best practices, a lack of consensus exists as to the exact definition of these terms. The World Tourism Organization (WTO; 1998, 8) defines sustainable tourism as "tourism that meets the needs of present tourists and host regions while protecting and enhancing opportunities for the future of the sector." This definition suggests that the overriding motivation is to maintain market share to ensure the continued financial success of the tourism sector with less emphasis on a triple bottom line of environment, society, and economy or people, planet, and profit, as promoted by sustainability advocates.

The confusion that exists regarding the concepts of sustainable tourism and ecotourism often results in public officials and the tourism sector using these terms interchangeably. Ecotourism is defined more narrowly because not all types of sustainable tourism can be considered ecotourism (see fig. 9.1). WTO has concluded that sustainable tourism guidelines and management practices that can be applied to all forms of tourism in all types of destinations are needed (1998). Hunter and Green (1995) developed the following list of criteria for sustainable tourism:

- Follow ethical principles that respect the culture and environment of the area, the economy and traditional way of life, and the political patterns.
- Involve the local population, proceed only with their approval, and provide for a degree of local control.

Ecotourism – A niche market promoting education, low impact, and respectful interaction between people and places.

Sustainable tourism – The overarching effort to make tourism ecologically and culturally sensitive and profitable.

Mass tourism – Tourism involving large numbers of people requiring extensive public and private infrastructure.

FIGURE 9.1. The relationships among tourism, sustainable tourism, and ecotourism.

- Keep intragenerational equity in mind, including fair distribution of benefits and costs.
- Plan and manage tourism with regard for the protection of the natural environment for future generations.
- Plan in a manner integrated with other economic sectors.
- Continuously evaluate impacts and initiate action to counter any negative effects.

An ever-present concern is "greenwashing," the appearance of being environmentally benign to maintain or attract market share when in fact these claims are false. Rider (2009, 15) defines greenwashing as "companies, products, or processes that are promoted as 'green' but may not actually adhere to good environmental practices." Greenwashing dilutes the importance and integrity of best practices and undermines consumer confidence in such programs, dissuading businesses from participation and ultimately not preserving the environment.

Ecotourism emerged in the late 1970s as mass tourism became the norm and people began to realize that this type of tourism isolated them from host cultures and often led to environmental degradation. As a niche market, ecotourism is generally expected to incorporate an environmentally friendly and culturally respectful approach. Some people dismiss ecotourism as a luxury available only to wealthy travelers seeking unique outdoor adventures, whereas others propose that any nature-based activity can be classified as ecotourism. This confusion about the definition of ecotourism presents challenges relative to its practice and evaluation. Honey (2002) identified eight elements of authentic ecotourism:

- Travel to natural areas
- Minimize impacts
- Build environmental and cultural awareness for hosts and guests
- Provide direct financial benefits for conservation
- Provide financial benefits and empowerment for local communities
- Respect local culture
- Be sensitive to the host country's political and social conditions
- Support human rights and international labor agreements

Carrying capacity, which is the number of visitors that a location can host without being degraded, is central to sustainable tourism and ecotourism. Carrying capacity is difficult to measure, particularly since capacity may be measured in cultural, social, and/or ecological terms. Clearly, ecotourism differs from sustainable tourism because it involves travel to natural areas, which generally means small- or medium-size tourism businesses are involved, to ensure that impacts on natural areas are minimized. Sustainable tourism, on the other hand, can involve large businesses that adhere to various sustainability criteria such as those in the list developed by Hunter and Green (1995).

While no statewide nongovernmental organization has a mission to further the goals of sustainable tourism in the Hawaiian Islands, the Hawai'i Ecotourism Association (HEA) was formed as an outcome of the statewide Conference on Ecotourism held in Waikiki in October 1994. The conference planning committee members continued to meet throughout 1995 to form HEA and quickly added over one hundred members. HEA is committed to diversifying Hawai'i's tourism industry through advocacy for and promotion of ecotourism as a means to protect the island's ecosystems and cultural landscapes. However, the organization's membership is primarily composed of two groups: small business owners/operators who have little time outside of the demands of managing a business to devote much time to HEA efforts, and sustainable tourism supporters who generally advocate for environmental conservation and cultural heritage regardless of their connections to sectors other than (eco)tourism businesses. Because of resource constraints and the lack of common goals between the two groups, HEA faces challenges in melding these disparate groups together to accomplish its mission.

The Hawai'i Tourism Authority (HTA) is the leading state agency for tourism whose responsibilities include setting policy, implementing marketing efforts, and sustaining a healthy visitor industry. HTA is attached to the Hawai'i Department of Business, Economic Development and Tourism

(DBEDT). In a report to the legislature in 2002, DBEDT acknowledged that the question of carrying capacity has been a concern for decades: "Clearly there is no simple answer to the question" and in developing a sustainable tourism planning tool, DBEDT recognized that the goal is not to "find a predetermined maximum number of visitors, but a strategy to ensure that growth is managed in a way that the quality of life of residents and the quality of the visitor experience remains high" (2002, 4). The following year's report noted that some public comments suggested no or limited future tourism growth and preferred economic diversification away from tourism or "new forms of tourism that take visitors out of the traditional 'big box' hotels and resorts: e.g., ecotourism" (2003, 22).

Shortly thereafter, HTA proposed the following definition for ecotourism: "Ecotourism in Hawai'i is an economically, socially and environmentally sustainable activity that responsibly and authentically connects visitors with Hawai'i's natural and cultural landscapes, resulting in beneficial exchanges among these landscapes, the host community, and the visitor" (2008). Recently, DBEDT's director suggested that "tourism has essentially remained stagnant for the last 20 years and can no longer be relied on to move the economy into a prosperous future" (Nakaso 2010). Cited as one of the goals for growth is investment to improve state lands, including protected areas. Although the objectives of the improvements would be varied, any such effort to improve state park infrastructure and habitat restoration on state lands would directly benefit ecotourism, given that protected areas are primary destinations of ecotourism operators, as well as outdoor recreation sites for residents.

THE SUSTAINABLE TOURISM MARKET IN HAWAI'I

In order to better understand visitor perceptions and expectations of ecotourism in Hawai'i, a class of University of Hawai'i at Manoa students surveyed over 200 visitors at several destinations on the islands of O'ahu and Maui between April and June 2010. The purpose of the survey was to learn more about visitor awareness of and preferences for ecotourism activities. The selected survey locations represented a variety of destinations and accounted for various activities, including marine and forest ecosystems and urban and rural sites. Surveys were completed on the streets of Waikiki ($n = 57$); at the trailhead to Manoa Falls ($n = 41$), a well-known hike for visitors and residents; at the Kapi'olani Community College Farmers' Market ($n = 53$), located on the slopes of Diamond Head; and at Waikiki Aquarium ($n = 79$). Surveys were fielded on the island of Maui at Waianapanapa State Park ($n = 26$) and in the town of Hana ($n = 19$), a gateway community to Haleakalā National Park that is visited by over one million people a year.

A review of those surveyed indicates 47 percent males, 46 percent Caucasian, 27 percent Asian, and 63 percent between eighteen and thirty-five years of age. Seventy-two percent had attended college, and while the respondents were relatively young, their incomes were relatively high, with 72 percent reporting annual incomes between $30,000 and $100,000.

The demographic information of respondents was compared with the visitor statistics compiled by HTA to determine if survey results are representative. In 2009, males accounted for slightly less than half of all visitors to Hawai'i from all countries of origin except Europe and Asian countries other than Japan. Young females represent a larger visitor base from Japan than from any other country of origin; thus, the gender composition of the survey results presented here is representative. The ethnic background of visitors in 2009 is not reported, nor is income; however, the West Coast of North America and Japan are the largest visitor markets for Hawai'i, and these markets would be expected to supply visitors who are likely to identify themselves as Caucasian or Asian. Visitors in 2009 were mostly in the age range of forty-one to fifty-nine years, except for the Japanese. In summary, survey respondents appear to be representative of Hawai'i's visitor population, except that survey respondents tend to be slightly younger than the visitors to Hawai'i overall.

The results indicated that 45 percent of the respondents were familiar with ecotourism and 64 percent considered Hawai'i as an ecotourism destination. While 63, 43, and 35 percent of respondents considered sustainability in accommodations, recreational activities, and food choices, respectively, while traveling, only 39 percent considered themselves as ecotourists, although 43 percent of the respondents were willing to donate time or money to further ecotourism objectives. These responses reinforce the conclusion that a lack of clarity about the role and definition of ecotourism exists. They also highlight the visitors' interest in supporting sustainability, particularly in accommodations.

Eighty-six percent of survey respondents reported that they were satisfied or very satisfied with the experience at the site where the survey was conducted, and 96 percent would recommend the sites to other people. Fifty-one percent of the respondents use the Internet to make travel plans, highlighting that online resources are a point of contact with visitors to promote operators adopting sustainability practices.

Use of the Internet by visitors to search tourism destination and activity information suggested the rationale for an additional study. A content analysis of self-identified ecotourism operator websites was conducted in July 2010 to evaluate how operators are portraying themselves on their business websites. As evidenced in the visitors' survey, the Internet is an important source of travel information. An earlier content analysis of media representations of ecotourism concluded that "policies are needed to govern ecotourism development in order to assure it is sustainable. . . . Hawai'i will continue to face conflicts as community members and outsiders move to profit from interest in ecotourism experiences" (Cusick, McClure, and Cox 2009, 33). The threat of greenwashing is more than false advertising; it puts into question the validity of commitments being made in support of sustainability practices.

The survey reviewed businesses on each of the main Hawaiian Islands that were found using the following keyword searches: ecotourism, sustainable tourism, authentic, cultural tourism, and heritage, as well as the names of individual islands and the Hawaiian Islands. The objective was to review sites that visitors would likely find doing their own keyword search of ecotourism-related businesses in the state of Hawai'i. All seventy-eight websites identified by the keyword searches were included in the study: eighteen on Kaua'i, ten on O'ahu, one on Lana'i, five on Moloka'i, twenty on Maui, and twenty-four on the island of Hawai'i. The website content analysis searched for the following information:

1. Memberships in local, regional, and/or international ecotourism associations and hyperlinks to those and/or other environmental organizations
2. A mission statement promoting ecotourism objectives and practices and demonstrated evidence of community participation and/or contribution of time and money
3. Staff biographies, including level and type of experience and expertise in natural and cultural history
4. Amount of information provided about people and places visited by the operator

Activities offered by the operators were categorized and allowed for multiple activities per website (fig. 9.2). The most popular activities offered were hiking (14%), snorkeling (13%), and kayaking (10%), indicating demand for both marine and terrestrial tourism activities. Over half of the operators (52%) offered marine activities, and nearly half (48%) offered terrestrial activities. Some operators offered activities not generally associated with ecotourism, such

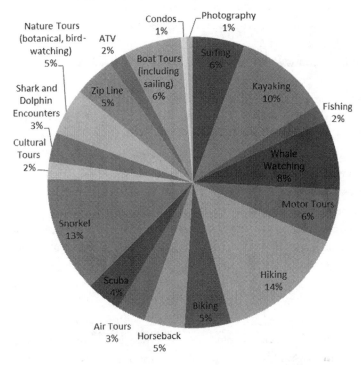

Activities Offered on Websites

FIGURE 9.2. Activities offered on surveyed, self-identified ecotourism websites.

as airplane/helicopter flights and all-terrain-vehicle tours, indicating that the use of ecotourism is being used as a keyword for marketing purposes. This supports the common misconception that any outdoor activity, even those that put tourists in contact with nature from the inside of a helicopter, is considered "ecofriendly."

SYSTEMS TO ASSESS SUSTAINABLE TOURISM IN HAWAI'I

Several benefits accrue to the tourism sector for participation in sustainability initiatives, including cost savings attributed to energy and water conservation, public relations and marketing benefits, public–private partnerships and subsidies for participation in the form of tax and/or product use incentives, and social networking. The challenges of achieving measurable levels of sustainability are significant, and the metrics for sustainability initiatives may require a comprehensive certification system to ensure the authenticity of green business practices to avoid greenwashing. The WTO concluded that sustainable tourism guidelines and management practices are needed to which can be applied to all forms of tourism in all types of destinations (World Tourism Organization 1998).

While a variety of assessment systems exist worldwide relative to quality and health, hygiene, and safety conditions for the tourism sector, conformity assessment systems for sustainability are relatively new. Certification programs are commonly used to assess sustainability. They involve industry benchmarks, standards, or guidelines, and they can improve efficiencies and overall performance (Scanlon 2007). In 2004, around one hundred programs worldwide certified tourism businesses for sustainability, with 78 percent of the businesses based in Europe and 68 percent focusing on accommodations. Programs in Australia and Costa Rica were considered in 2005 to be models of effective certification that furthered the goals of a socially responsible tourism industry. Since 2005, Green Globe, Sustainable Travel International, and Partnership for Global Sustainable Tourism, among others, have worked to develop more comprehensive programs that are accepted worldwide (Cox et al. 2008).

According to Rider (2009), the goal of a third-party green certification system is not to be the standard for every situation but instead to create a paradigm shift away from the status quo toward innovative leaders that encourage others to join in and advocate for a greener tourist sector. "The overall aim is that the label of [a] certification program will be recognized by consumers or distribution channels, and considered as added value that leads to its acceptance in the marketplace, to support the marketing of companies that meet standards" (Font 2007, 387). Honey (2002) concludes that a certification program should have the following elements:

- Standards
- Assessment
- Certification
- Accreditation
- Recognition
- Acceptance

If all of these elements exist, then the costs of a certification program should, in the long run, be borne by the tourism sector and its customers who view it as an efficient, effective means of achieving a productive relationship.

Hawai'i started down the path towards a sustainability assessment system with the passage of Act 133 in 1976 (see Mak 2008, 55). Act 133 "required a plan to guide future tourism development that would meet the needs of both tourists and residents and at the same time protect the environmental, social, and cultural resources of the state. In sum, the act directed the state DBEDT to craft a plan for 'sustainable tourism'—a concept that would not be defined until 1992." In 1998, Act 156 created HTA with one responsibility being to develop alternative forms of tourism, including "agri-, cultural, edu-, health and wellness, eco-, and techno-tourism" (Mak 2008, 65). The HTA spent $1.2 million on a Sustainable Tourism Plan that was the product of a nearly two year process of bringing stakeholders together to define a vision, set goals, and identify indicators (Sheldon, Knox, and Lowry 2005).

The socioeconomic commitment, however, needs to be scaled up from individuals to institutions and from policy statements to actual practices. Sustainability of the scale required to transition an isolated island archipelago requires significant investments of time, labor and capital, and ingenuity. "The challenge to sustainable [tourism] management is to keep those unique elements that are part of the mystique that attracts tourists. Shifts toward homogeneity of tourism experiences without giving the visitor an experience of 'sense of place' will not be sustainable in the long run for the host community or for the tourism markets" (Sheldon 2005, 10). Nonetheless, "as an organizing principle, sustainability reflects the fundamental relationships that underlie ecological, economic, and social concerns. It offers the possibility of bringing social change values into the mainstream and pushing the mainstream toward sustainable practices" (Edwards 2005, 133). The dependence of Hawai'i's tourism sector on the mass market and large scale of tourism infrastructure throughout the islands challenges policy makers and the industry to achieve the legislative goals of the Hawai'i 2050 sustainability plan, which, among other benchmarks, calls for the development of a more diverse and resilient economy (Cox et al. 2008; Social Sciences Public Policy Center 2010). Development of a sustainability strategy and a means of assessing its progress, such as a certification program, are needed.

Efforts to change management practices in the tourism sector have occurred recently in the state. Some of the change in practices may be attributable to the realization in the tourism sector that visitors are interested in supporting sustainable enterprises with their travel dollars. Regardless of the reason for these recent successes, each effort is discussed in the following sections.

DBEDT and the Hawai'i Department of Health (DOH) developed the Hawai'i Green Business Program to encourage business participation without the expense generally associated with private certification programs. The initial focus relative to the tourism sector was on hotels because they are among the largest consumers of electricity and electrical system efficiencies would immediately create cost savings. LEED (Leadership in Energy and Environmental Design) and Green Globe,[1] briefly presented here, are two of the sample certification programs reviewed in the process of developing the program because they focus on the built environment.

The US Green Building Council developed LEED, perhaps the best-known and most rigorous green building certification system, in 2000. LEED for Existing Buildings (LEED-EB) was analyzed since DBEDT's program focuses on creating conditions that result in a quick return on investment through efficiencies and conservation measures. LEED-EB has seven categories in the certification system: sustainable sites, water efficiency, energy and atmosphere, materials and resources, indoor environmental quality, innovation in operations, and regional priority credits.

Green Globe is an international certification program that promotes best practices to address what both the WTO and the World Travel and Tourism Council (1999) consider as problematic areas in the tourism industry. The program's criteria target primarily energy consumption, solid waste management, potable water use, and associated retrofits resulting in cost savings that appeal to the business community, as well as land use planning and sustainable designs of the built environment. Broad stakeholder involvement is emphasized to develop sustainability policies and identify benchmarks for their environmental performance, while encouraging completion of an independent audit to ensure compliance with industry best practices.

Green Globe 21 has a checklist of criteria that businesses must achieve before being considered for certification. The process can cost several thousand dollars, excluding the cost of site inspections, and take as long as two years, depending on the size of the company. Certification has to be renewed every two years. There are four levels of Green Globe 21 certification: bronze, silver, gold, and platinum. Once certified continuously for five years, the Green Globe logo may be used for marketing purposes.

Overall, the objectives of the Hawai'i Green Business Program are to recognize businesses that implement a checklist of green initiatives similar to LEED and Green Globe 21 criteria. Established in 2004, the program has recognized twenty-five hotels and resorts in the state of Hawaiʻi and expanded to other sectors such as the office/retail, government services providers, and restaurant/food businesses. Participation is intended to improve Hawaiʻi's image as environmentally conscious, while being business friendly, with a primary focus to reduce energy and water consumption and greenhouse gas emissions and solid waste; recycling as much as possible to minimize the waste stream going to landfills; ensuring respect for host culture values and cultivation of a sense of place; and education of employees and customers regarding sustainability best practices. To address potential participant concerns about costs, DBEDT's green certification program is free and designed to nurture the necessary paradigm shift towards sustainability. The self-assessment process used by the program is not considered as rigorous as a third-party certification (Honey 2002).

The hotels participating in the Hawaiʻi Green Business Program have reduced the amount of resources used by staff and visitors, but equally important, hotel management and staff developed internal programs to sustain program objectives (table 9.1). As table 9.1 indicates, the outcomes vary by hotel. Many of the reported results can easily be translated into decreases in costs due to reductions in resource use. However, the diverse nature of these outcomes makes sustainability progress hard to track and reinforces the need for a consistent policy relative to the state's sustainability plan. Travel agencies and media have also recognized the efforts of some hotels, which create additional incentives to join green certification programs.

HEA Ecotour Certification Program

In 2001, with grant support from the HTA, the HEA worked with several other organizations to explore the creation of a certification program to ensure the authenticity and accuracy of cultural information shared in tours and other activities involving visitors. Workshops were conducted on Oʻahu and Maui, and the conclusion among participants was that Hawaiʻi was not ready for a formal certification program, because of a lack of sustained funding for such an effort.

HEA partially funded a University of Hawaiʻi at Manoa graduate student in geography to complete a survey of its members to explore their interest in an ecotour certification program (Bauckham 2005). The thesis concluded that members supported a certification program as long as it was simple and inexpensive. Members hoped that consumers would be willing to pay more for

TABLE 9.1. Examples of results from hotel greening initiatives

---

Hyatt Regency Maui
Recycling savings:
- 3.1 tons of plastic
- 24 tons of cardboard
- 29 tons of glass
- 1 ton of mixed paper
- 3.35 tons of aluminum
- >10,000 gallons of fryer oil (converted to biodiesel)

Hyatt Regency Waikiki
- Installed 1,400 LED lights
- Reduced electricity usage by 287,554 kilowatts
- Installed 0.5 gallon/min aerators in guestroom faucets
Recycling savings:
- 140 tons of cardboard
- 40 tons of glass
- 450 tons of food

Moana Surfrider Waikiki
- In-room paper, glass, linen, and can recycling
- Installed over one-third indigenous plants for exterior landscaping
- Installed heat pumps and exchangers to recover waste heat for hot water

Sheraton Hotels Waikiki
- Recycling program at four Waikiki properties
- 55% waste recycling
- Saved $100,000/year
- Sheraton Princess Kaiulani is Energy Star–labeled building
- Starwood is an Energy Star Partner
- Sheraton Waikiki is a Green Business Awardee

Hilton Waikoloa Village
- Na Lima Hana CARE Committee: programs to raise awareness of environmental responsibility
- 202 community service/school programs for 5,420 participants (2005)
- LEED Silver-certified building

Waikiki Marriott
- $350,000 in annual energy savings
- $220,000 in rebates
- Replaced old air conditioning equipment with new chillers
- Installed energy-efficient lighting

Hawai'i Prince Hotel Waikiki
- Retrofitted guestroom lighting: estimated annual savings of $92,901.88 @ $0.13/kilowatt-hour
- $17,475.00 in rebates
- Uses saline pool technology to cost-effectively sanitize pool

Mauna Lani Kohala
- Largest resort photovoltaic system in world: ~500 kilowatts
- Rated by Condé Nast as one of the top three earth-friendly getaways
- Composts green waste: 5 tons/day

Hilton Hawaiian Village
- Water use down 35% per room
- Over 33 million gallons saved
- Low-flush, low-flow fixtures
- Reuses air conditioning cooling water
- Energy management system
- $400,000 annual savings
- Recycle waste: 1,300 ton/year

Maui Prince Hotel Makena Resort
- Created a special Hui Oma'oma'o (Green Team)
- Reduced water consumption >10 million gallons in 2008
- Recycles over 50% of waste stream

---

certified ecotourism products and/or services and that certification would promote ecotourism across the state. As a result of these conclusions, an HEA committee developed a review process for ecotourism operators based on the descriptions of other certification programs in use around the world. The review is a self-assessment that requires an application fee between $50 and $350, depending on the total number of employees. Only five operators participated in the review process, and none of them promote the review as part of their online marketing strategy. Thus, the program lacks statewide participation due, in part, to a lack of resources, which are needed to educate visitors, residents, and the tourism sector about the program and its benefits. The HEA's review process for ecotour operators has not been as successful as the Hawai'i Green Business Program, so the process had very little acceptance.

In order to encourage more buy-in within the tourism sector, HEA received a small grant in 2009 from HTA to help move the review process forward to an ecotour certification process. Given the difficulties with developing an effective certification program, the grant's limited resources hampered efforts to make significant progress. By September 15, 2011, fourteen ecotour operators were certified as part of this pilot effort. Public meetings held across the state about the program supported the conclusions of the visitor survey presented above that people do not understand what ecotourism is, while more agreement exists about what the word "sustainability" means and what sustainability looks like. Another concern is that the word "ecotourism" generally means that the tourism business is small to medium in size and excludes various activities. A certification program in Hawai'i is therefore more likely to have an impact if the focus is sustainability certification rather than ecotourism certification. An ecotourism certification program may prove to be too restrictive and not facilitate a large amount of progress toward a more sustainable Hawai'i by 2050.

Maintaining Hawaiian culture is very important. Most other certification programs do not place such emphasis on one culture, which means that Hawai'i should consider maintaining unique certification standards because of the importance of maintaining a Hawaiian sense of place. At the same time, nearly all of the ecotour operations in the pilot program were staffed with a large percentage of people that moved to Hawai'i as adults. Ecotour operators in the pilot program generally do not feel that the state is particularly attentive to their needs because local residents are not supportive of the tourism sector. Residents need an overall understanding of the tourism sector's value so that they can seize the niche opportunities that tourism offers. At the same time, residents can use certification programs as a means to interface with the tourism sector to address their concerns.

West Hawai'i Voluntary Standards

The Coral Reef Alliance (CORAL) began developing the West Hawai'i Voluntary Standards for Marine Tourism in 2008, and by the summer of 2009 the community members on the western side of the island of Hawaiʻi had voted to accept them. The standards address scuba diving and snorkeling; general boating, including kayaking and surf schools; wildlife interactions, including dolphins, whales, monk seals, turtles, manta rays, sharks, and invertebrates; and shoreline activities. Each set of standards outlines "environmental performance, conservation practices and operational safety" for the relevant type of marine tourism (Coral Reef Alliance, n.d.).

CORAL promulgates these standards by entering into a memorandum of understanding with a marine tourism business, and by June 2011 thirty businesses in West Hawaiʻi agreed to work with CORAL to evaluate the "overall effectiveness, attainability, and affordability" of the standards. CORAL's work with marine tourism businesses is more complex than the review process developed by HEA. Marine ecotourism operators agree to "allow, support and facilitate the assessment of these standards through the following methods: (1) passenger/client exit surveys; (2) self-assessment; (3) peer review; and (4) anonymous third-party community participation and feedback" (Coral Reef Alliance, n.d.). CORAL also assists businesses in implementing the standards by developing and disseminating various educational tools and materials while also providing specialized training in sustainable marine recreation to businesses.

CONCLUSION

This chapter identified Hawaiʻi's challenges and accomplishments in the area of sustainable tourism. Larger businesses in the tourism sector, such as hotels, are making progress in adopting sustainability practices because they are able to realize cost savings from reducing their resource use and can benefit from the positive publicity that is thought to be associated with such efforts. Efforts to reduce the resource use by hotels are also receiving attention from government organizations and nonprofit organizations across the state. Since the majority of certification programs worldwide focus on accommodations, the most efficient way to start moving toward ensuring compliance likely lies in this portion of the tourism sector.

Ecotourism operators, however, are generally much smaller and provide recreational services to visitors, which are much more difficult to evaluate for sustainability. At the same time, small operators may need specialized assistance to reach certification standards for services. The success of the West Hawaiʻi Voluntary Standards for Marine Tourism indicates that a certification program simultaneously providing education to operators and their staff may be crucial. Small operators may lack the human and financial resources required to become more sustainable and remain competitive with other businesses that do not pursue sustainability practices. Clearly, more resources have been allocated in Hawai'i to assist small marine operators than land-based operators. Federal resources appear to have been a significant source of funds for the marine-based efforts.

Ecotourism businesses also tend to be more integrated into the local community than those businesses that focus on serving the mass market, such as resorts, simply because of their size. They are also subject to larger risks and have more fragile profit margins than these larger businesses, which puts them into a more tenuous economic situation. Residents may, in fact, prefer that visitors stay isolated in areas such as stand-alone resorts in order to lessen the impact of visitors on their lives. Residents who experience the negative impacts of tourism, such as high resource costs, environmental damage, and societal changes may never interact with mass tourism participants and miss the rich, positive interaction that can result. Residents employed by the tourism sector may be perceived as the only people in the community who benefit from tourism. At the same time, residents may not fully understand that economic diversification is difficult to achieve on an isolated island archipelago. Residents may be reluctant to commit any resources to tourism, while at the same time, finding a competitive advantage for Hawai'i that can match or exceed tourism will be challenging. A commitment of resources by policy makers and residents is likely to be needed in order to support sustainability efforts and to support businesses that are not part of the mass tourism market. Any lack of commitment will be felt by small businesses first, and these same businesses will be needed to ensure that the visitors' experience is authentic.

At the same time, Hawai'i's visitors may not fully understand what sustainability and ecotourism entail, if the visitor survey results discussed in this chapter accurately reflect their knowledge. More education is needed to ensure that visitors can understand how to support sustainable tourism once they arrive in the islands. While the website survey revealed that online information sources are not particularly linked to sustainability, many visitors are getting their information from websites. This provides a window of opportunity to further promote sustainability practices by the tourism sector.

Increasing competition from other destinations is fueling interest in alternative forms of tourism. Currently, more attention as to how to manage these forms of tourism is needed. For example, operators from other states are now leading hiking groups, while businesses that have existed in the state for many years, employ a local workforce, and contribute to the community in a variety of ways may find it difficult to stay in business due to the operating conditions they face. The state has a limited set of policies for supporting these small-scale businesses and few resources to monitor and enforce what few regulations exist. Commitment to a political process that includes all members of the local community is needed. A process that promotes openness such that all persons who will be affected be allowed to participate; balance so that no single interest category can account for more than one-third of the group deciding on the outcome; and a degree of consensus such that all views and objections be considered and an effort be made to resolve differences is needed.

The state of Hawai'i is a recognized world leader in tourism development, serving as a planning model for other isolated destinations, whether Pacific Islands or elsewhere. This planning role also provides an opportunity to demonstrate sustainability best practices, given the vulnerability of the islands to energy and food insecurity, environmental degradation, and climate change. Many shades of green exist in Hawai'i, and more action is needed to ensure that the state's tourism sector is on the path to achieve sustainability by 2050.

ACKNOWLEDGMENTS

The authors share equal credit for the authorship and acknowledge the contributions of Miguel Die-Gonzalez, Bixler McClure, Brenna St. Onge, and Gail Suzuki-Jones.

NOTE

1. Green Globe is a global certification for sustainable tourism. This is not to be confused with Green Globes, which is a system for building environmental design and management.

REFERENCES

Bauckham, W. 2005. *Putting Traffic Lights on the Road Less Traveled: Ecotourism Certification and Its Potential for Hawai'i*. Master's thesis. Honolulu: University of Hawai'i at Manoa.

Coral Reef Alliance. n.d. "Voluntary Standards for Marine Tourism in Hawai'i." http://www .coral.org/west_hawaii_standards.

Cox, L., M. Saucier, J. Cusick, H. Richins, and B. McClure. 2008. "Achieving Sustainable Tourism in Hawai'i Using a Sustainability Evaluation System." Honolulu: University of Hawai'i at Manoa.

Cusick, J., B. McClure, and L. Cox. 2009. "Representations of Ecotourism in the Hawaiian Islands: A Content Analysis of Local Media." *Journal of Ecotourism* 9 (1): 21–35.

Edwards, A. 2005. *The Sustainability Revolution: Portrait of a Paradigm Shift*. Gabriola Island, BC, Canada: New Society Publishers.

Fennell, D. 2001. "A Content Analysis of Ecotourism Definitions." *Current Issues in Tourism* 4 (5): 403–421.

Font, X. 2007. "Ecotourism Certification: Potential and Challenges." In *Critical Issues in Ecotourism: Understanding a Complex Tourism Phenomenon,* edited by J. Higham, 386–405. Oxford: Elsevier.

Honey, M. 2002. *Ecotourism and Certification: Setting Standards in Practice*. Washington, DC: Island Press.

Hawai'i Department of Business, Economic Development and Tourism (DBEDT). 2002. *Planning for Sustainable Tourism in Hawai'i*. Honolulu: State of Hawai'i.

———. 2003. *Planning for Sustainable Tourism in Hawai'i: A Study on the Carrying Capacity for Tourism*. Honolulu: State of Hawai'i.

Hawai'i Tourism Authority. 2008. *HTA's Sustainable Tourism Strategy*. http://www.hawaii.edu/envctr/ecotourism/docs/HTASustTourism2608.pdf.

Hunter, C., and H. Green. 1995. *Tourism and the Environment: A Sustainable Relationship?* New York: Routledge.

Mak, J. 2008. *Developing a Dream Destination: Tourism and Tourism Policy Planning in Hawai'i.* Honolulu: University of Hawai'i Press.

Nakaso, D. 2010. "Future of Tourism Called into Question." *Honolulu Star-Advertiser,* June 3.

Rider, T. 2009. *Understanding Green Building Guidelines: For Students and Young Professionals*. New York: Norton.

Scanlon, N. L. 2007. "An Analysis and Assessment of Environmental Operating Practices in Hotel and Resort Properties." *International Journal of Hospitality Management* 26: 711–723.

Schaefers, A. 2011. "Tourism Outlook: Up? Down? (It Depends on Whom You Ask)." *Honolulu Star-Advertiser,* May 20.

Sheldon, P. 2005. "The Challenges to Sustainability in Island Tourism." Occasional Paper 2005-01, School of Travel Industry Management, University of Hawai'i. http://www.tim.hawaii.edu/ctps/Sheldon_Challenges_to_Sustainability.pdf.

Sheldon, P., J. Knox, and K. Lowry. 2005. "Sustainability in a Mature Mass-Tourism Destination: The Case of Hawai'i." *Tourism International Review* 9 (11): 47–59.

Social Sciences Public Policy Center. 2010. *Hawai'i 2050 Update*. Honolulu: University of Hawai'i at Manoa.

United Nations. 1987. *Report on the World Commission on Environment and Development: Our Common Future*. http://www.un-documents.net/our-common-future.pdf.

World Tourism Organization (WTO). 1998. *Tourism Services: Background Note by the Secretariat. World Trade Organization: Council for Trade Services*.

World Travel and Tourism Council. 1999. *WTTC Hawai'i Tourism Report 1999: How Travel and Tourism Affect Hawai'i's Economy*. http://Hawai'i.gov/dbedt/info/visitor-stats/econ-impact/WTTC99.pdf.

TEN

Successful Sustainability Movements in Higher Education

SHANAH TREVENNA

We can't solve problems by using the same kind of thinking
we used when we created them.

—*ALBERT EINSTEIN*

For anyone who scratches beyond the surface of the typical tourist's experience, Hawai'i has many surprises in store. The Native Hawaiian influence and deep community fabric permeate all aspects of the culture. Most conferences and public events begin with a Hawaiian prayer or chant, and when meetings end, many participants hug each other warmly. The deep and intricate interconnections of people and organizations lead to only a few degrees of separation. There is also an undying spirit of community, belief, vision, and support. Among many diverse stakeholders, there is surprising unity in a shared vision and desire for energy and food self-sufficiency, minimization of waste, conservation of water and natural resources, and agricultural diversification for a strong, local economy. By actualizing this potential, Hawai'i would provide a model for positive change for the world, with the University of Hawai'i (UH) leading the way.

The majority of today's college students desire the presence of sustainability in all areas of campus operations and programs when selecting a college. A 2010 Princeton Review survey of over 8,000 college applicants across fifty states found that 65 percent of respondents said that a college's commitment to sustainability would contribute to their decision. Research also shows that, for both teachers and students, integrating sustainability comprehensively through multiple paths, including core requirements, degree requirements, extracurricular programs, interdepartmental minors, and mission statements instills attitudes and skills required to be positive change agents for society (Rowe 2002).

Change agents, or leaders who think in new ways, communicate, mediate, teach, coach, set goals, and achieve them, are essential to the evolution of the institutions and organizations that shape life in Hawai'i. In the context of organizational development, the role of a change agent is to change the worldviews of stakeholders towards an attractive, but unknown, future, as well as to support the management of change towards that future (Coskun 2008, 2). Higher education has historically played the role of producing knowledge rather than producing change; however, there is a great need for colleges to become "facilitators for dialogue"—change agents—rather than "repositories of truth" (Tierney 2001). Since technology has provided unprecedented universal access to information, the academic claim to isolated spheres of knowledge domains no longer exists. Instead, colleges need to better prepare students to become experts in the ability to cross intellectual, geographic, social, and cultural borders (Tierney 2001).

Colleges that embrace this new mission still face the challenge of *how* to implement this level of institutional change. The individuals and parties commonly tasked with this monumental undertaking are campus administrations, faculty, or facility managers, yet these stakeholders often work in isolation, and students are left out of the equation. This chapter shares the case study of a student led movement at UH that created a tangible proof of concept of student empowerment that engaged all stakeholders in creating sustainable, institutionalized systemic change.

IMPLEMENTING INSTITUTIONAL CHANGE

When considering which stakeholders on campus are ideally positioned to lead change, students, who are often left out of planning processes, are actually in the best position to lead change. Urban planning theorist Leonie Sandercock (1998, 102) recommends that leadership to overcome the tensions between those in power can come from marginalized groups such as women and people of color who have experienced marginality, exploitation, and domination. From her perspective, the disenfranchised that are not politically or economically entrenched and have built tools to deal with oppressive processes can be the ones to create a common ground for all stakeholders. The example of Sustainable Saunders at the University of Hawai'i at Manoa (UHM) provides insight into how one group of students found a focal point for their efforts, came to understand what suffocates change on campus, and developed tools and strategies for addressing resistance to make a campus and statewide sustainability movement in education.

As is the case at many institutions of higher education, sustainability at UH began in various pockets of campus life, driven by passionate leaders doing everything they could do within their scope of influence. The classes, departments, or projects lacked system-wide support and coordination and often lost stamina as they hit barriers to expansion. At UHM, sustainability was constrained, overseen by one faculty member. Campus culture had accepted that the Office of Sustainability was off the beaten path, which bolstered the perception that sustainability was being taken care of by "others" and not core or integrated into the mainstream campus experience. When the Sustainable Saunders project began in 2006, UHM was similar to many campuses—there was a disconnect between top-down efforts by the facilities department and the bottom-up needs of the academic community. A common ground was needed to provide a tangible opportunity to connect, face opposition, and collaborate.

The opportunity to create a common ground presented itself at UHM with the creation of the Sustainable Saunders initiative. Saunders Hall was a typical building within the UH system in that it was several decades old and had some original and retrofitted lighting and air conditioning. Built in 1974, each of its seven stories houses a different department within the College of Social Sciences. In 2006 the dean of social sciences, Legislatures in Residence, Public Policy Center, and Office of Sustainability designated Saunders Hall as the campus pilot project where new technologies and change-making projects could be tried. The projects that proved successful could be rolled out throughout the campus.

While it was not obvious to new students, a lot of groundwork had been laid on campus, and this initiative was just one example. Many stakeholders had been talking about the need for change, and the idea had been brewing for many years. They hoped Sustainable Saunders would provide the unprecedented opportunity to bridge the many isolated departments and also address the long-standing conflict between academia and facilities.

Following the creation of Sustainable Saunders, the UHM Chancellor's Office announced an energy summit. A campus leader demonstrated stakeholder engagement by bringing all of the players together, including academia, facilities, the local utility company, renewable energy companies, and mainland consultants, to explore how to address their energy bill. The result was UHM's first clean energy policy, which set the goal of a 30 percent reduction in campus energy use within six years.

Prior to the announcement of this policy, the various departments that needed to act had not been able to justify allotting time or resources to sustainability. After the creation of the overarching energy goals at the chancellor's summit, and people realized the project was in line with the chancellor's goals, those who did not originally have an interest in the project before became engaged in supporting it. The first lesson from Sustainable Saunders was that to launch systemic change, clear prioritization, direction, and measurable goals had to come from top-level leadership.[1]

CHANGE AGENTS UNITE

The planning profession has long called for a model that empowers the disenfranchised to dispel top-down and bottom-up tension by building a bridge between the two and uniting these forces into a power-sharing network for change (Sandercock 1998, 102). The students who worked with the academic community and facilities at UHM proved such a model was viable. A steering committee of eight students was assembled based on their authentic passion and their demonstrated ability to get involved and take positive action. These students possessed change-agent skills as articulated by College Student Educators International (n.d.): resiliency, curiosity, passion, self-awareness, tenaciousness, assertiveness, empathy, and optimism, among others. For example, one creative architecture student had installed large displays around campus that had silent but flashing warning lights that drew pedestrians closer to the building to investigate. The students named their group the HUB, which stands for Help Us Bridge, to demonstrate their intent to create excellent proposals and work with everyone on campus. The dynamic that emerged when these students came together created a unique team culture, which they eventually characterized according to the following core values.

Core Values

VALUE 1: SUSTAINABILITY IS SEXY: Sustainability as a vehicle for change is oriented towards what is gained, not lost; towards solutions, not problems; towards the positive, not the negative; towards sexy (exciting, a world of new opportunities, confidence, "beyond the norm"), not sacrifice (focusing on what you don't have).

VALUE 2: PULL, NOT PUSH: Teams thrive when members work on what they love and not what they feel forced to do. There should be an attraction to the areas of change that students want to work on, so they feel drawn to the work and have more energy after engaging in the work, rather than less.

VALUE 3: PERSPECTIVES, NOT AGENDAS: Change is a creative collaboration that comes from many disciplines and voices. No one needs to have it all figured out. It is not about knowing the answers or having individual agendas but about working with diverse perspectives to find solutions that work.

VALUE 4: PROCESS, NOT DESTINATION: There is value in just doing one's best rather than trying to be perfect. A step in the right direction reveals the next step. Overarching principles provide guidance but having fluidity drives the process.

VALUE 5: BEYOND "US" VERSUS "THEM": It is important to diffuse polarizations between opposing parties and viewpoints. Finding a common ground gets things done and forges creative solutions.

VALUE 6: CREATE BY CONSENSUS: It is important to respect voices of all ages, levels of interest, experience, and walks of life so the big picture can come together. Decision-making processes should use consensus and be open and transparent.

VALUE 7: CYCLIC, NOT LINEAR: With linear processes, outputs from a process are often wasteful in that they do not feed into another process. Cyclic processes involve returning those same outputs into the next process, so waste equals food or fuel. This applies to human interactions as well. The goal is to recognize linear situations and make them cyclic by taking the negatives and turning them into positives that help achieve a common ground.

A key to the success of this group was that they were an interdisciplinary team. Since the problems they wanted to address were beyond the scope of any one discipline, many perspectives were needed to answer the complex questions and address the broad issues characterized by sustainability (Klein 1990, 11). The original eight members of the Sustainable Saunders steering committee represented architecture, engineering, tourism management, indigenous medicine, environmental studies, and microbiology. The interdisciplinary nature of their team created an initiative that focused on broad, cross-disciplinary approaches.

It became clear from student feedback that the group's culture of meaningful interaction and optimal experience prompted them to remain engaged and inspired for years. Students who graduated made arrangements with their employers to continue attending HUB meetings. Years after the formation of the team, HUB alumni still consider themselves team members and manage the HUB social media outlets.

The psychology of optimal experience sheds light on the HUB experience. Psychologist Mihaly Csikszentmihalyi's book *Flow: The Psychology of Optimal Experience* (1990, xi) summarizes decades of research on optimal experience including joy, creativity, and the process of total involvement with life, which the author calls "flow." A flow activity is defined as providing a sense of discovery, a creative feeling of propelling the person into a new reality while transforming the self through growth (74). For an activity to create flow, it must have rules that require the learning of skills, set up goals, provide feedback, and make control possible (72). When activities have these attributes and are in line with someone's natural curiosity and interest, people can become completely absorbed and ultimately have an enjoyable, optimal experience.

This description of flow reflects the many positive aspects of the HUB culture. The students had autonomy over their contribution, as well as the support of the team. After they mastered certain skills, they looked for new challenges to stretch their abilities. The HUB grew their capabilities and skills to meet growing demands and evolving challenges. Students became change agents by being supported in cultivating their natural desire to learn and engage in tasks. They called it a "pull-not-push" philosophy, where they were encouraged to tap into what really "pulled" them and to not sign up for something if they felt "pushed." The diversity of the team was key to its success since nearly every task was accomplished through someone's pull. Because of the diversity of the group, the team knew that if they signed up for something they didn't really want to do, they were taking the opportunity away from a teammate who really wanted to do it. As surprising as it may seem, every task was accomplished through someone's "pull."

Whether they had the relevant skills or experience to accomplish the task they were pulled to do did not matter as long as they held a sincere interest. They were encouraged to engage their team members and to seek mentorship from others, but they were each responsible for their own self-directed learning. In the Hawaiian language, this kind of personal responsibility is called *kuleana.* It is often described as one's duty or responsibility. Whenever students were committed to a task from a sincere pull, it became a part of their *kuleana,* and they were motivated to see it through. They agreed to honestly communicate with the team if a task became too overwhelming or stressful. Stress was not supposed to be a natural part of what they did. If a student was feeling stressed, afraid, or overwhelmed, he or she was encouraged to seek support from other HUB members.

Another way to identify "pull" tasks is that the task energizes the participant. One example of a demotivator is a perfectionist approach, which HUB called the "perfectionism vortex." Since they were committed to doing a great job, students sometimes felt overwhelmed with possibilities. This syndrome was addressed by ensuring that all projects were performed in groups, even pairs. They would keep each other on track and provide support to each other.

Another technique for avoiding the "perfectionism vortex" was to brainstorm roadblocks, maintain direction as a group, and use meetings as think tanks. Many students cited HUB meetings as a weekly highlight: engaged and productive collaboration fostered creative synergy, leading to clear, exciting solutions. The HUB reached consensus for every challenge, no matter how complex. Thus, they ensured that meetings were for creative, collaborative problem solving rather than negativity.

LAUNCHING CHANGE

Many student groups seem like they act from a sense of entitlement since they want quick results due to their short time horizon on campus. The demand for quick action does not match the bureaucratic processes and the cultural norms of those who have been working within them for decades. The inability to recognize this friction in timeframes has been the fatal flaw of many campus groups striving for change.

Policy theorists have categorized different types of change processes described by Kelly et al. (2009, 15; see also table 10.1). The typical aggressive policy change advocated by students can be described as fast and cumulative, such as the "classic paradigmatic" approach. While this type of intense change may seem desirable, it may actually close off path-dependent policy pathways that may have the most transformative potential. An alternative option that reflects the HUB approach is "progressive incremental," which has a better opportunity for acceptance. Progressive increments in the form of slow, cumulative changes work best with both the student and campus timeframes.

TABLE 10.1. Types of change processes

| Directionality | Tempo | |
	Fast	Slow
Cumulative	Classic paradigmatic	Progressive incremental
In equilibrium	Faux paradigmatic	Classic incremental

While each increment is small enough to be accepted and implemented, the key is selecting increments that contribute to a cumulative total that moves towards the preferred future (Kelly et al. 2009, 14). "Progressive incremental" differs from the classic perception of incremental since the overall path continually moves towards overarching goals. This means that each increment, no matter how small, is of value if it moves irreversibly towards the end goal. If there is pushback from the community, or if they are so unhappy with the activity that they undo the increment, then nothing is accomplished.

IMPLEMENTING CHANGE

STEP 1: UNDERSTANDING THE TRIPLE BOTTOM LINE: As the HUB began to contemplate which projects were worth their effort, they needed criteria for assessment. How would they know if a project actually created a worthy change? Traditionally, many organizational decisions, from procurement to policy, are made from a single bottom line: money. But what if every decision also considered the people affected, including their culture, health, happiness, and fulfillment? And what if the environment were to also experience a positive result? Environment, society, and economy are often considered trade-offs. A framework called the triple bottom line strives to measure the success of initiatives not by economic factors alone but also by social and environmental benefit. If the decision reduces waste or resource use, improves conditions for society, *and* increases profits, the triple bottom line of sustainability is satisfied.

STEP 2: WASTE AUDIT AND RECYCLING: The HUB decided that their first project would be a waste audit, often called a "dumpster dive," on the front lawn of Saunders Hall, showing that they were not afraid to get their hands dirty. Next, they implemented recycling bins on every floor of the building, and through a subsequent "dive" found that they were able to divert 70 percent of plastic bottles from the landfill. They displayed the waste audit findings outside the elevators on the main floor so that all could see the results of their collective efforts. They knew their first project had to be highly visible and widely accessible so that everyone in Saunders Hall would know of their efforts were of service. HUB members knew this was a worthy project since people, the planet, and profits all simultaneously benefited.

STEP 3: ASSESS AN ENERGY-USE BASELINE: The next step was to determine how to reduce the energy bill for Saunders Hall. The HUB began by using an important axiom from engineering: "You can't manage what you don't measure." To create measurable goals, they first established a baseline that was a snapshot of the current metrics of interest. A local engineering firm generously performed a comprehensive energy audit. It showed that most of the building's energy was used for two functions: 42 percent for lighting and 46 percent for air conditioning.

While measuring the energy use provided an economic baseline in terms of the cost of electricity use, and an environmental baseline in terms of the amount of oil used and pollution produced, they needed to find a way to measure the social baseline to meet the triple bottom line. UHM's Public Policy Center worked with them to survey the community. Interestingly, the number one complaint in the survey was that the air conditioning was too cold, and the building occupants complained that the lights were too bright. They found that money was being spent that actually made people uncomfortable rather than more comfortable. By identifying community perceptions, they expanded the traditional engineering path for addressing what is often viewed as a straightforward technical challenge of reducing an energy bill.

STEP 4: SET AN ENERGY REDUCTION GOAL: For a goal to inspire change, it must result in a scenario beyond business as usual and be in line with the community's preferred future. Everyone should look at the goal and think, "I want that." The HUB set the ambitious goal of reducing Saunders Hall energy use in one year by 30 percent.

STEP 5: LIGHTING: THE "LOW-HANGING FRUIT": The HUB decided that their first attempt to incrementally reduce energy use would be to address the lighting, because reducing lighting is relatively simple and cost effective. While there were complaints that the lights were too bright, they knew they had to find a way to measure "too bright" so they could manage it. They discovered that the Illuminating Engineering Society recommended healthy standards for the amount of illumination at a task depending on the type of activity. The units are in foot-candles, which is the amount of illumination a birthday candle emits at a distance of one foot.

The Illuminating Engineering Society recommended that for typical reading and writing in a classroom, 30–50 foot-candles would be ideal. Using a Lutron LX-1010b light meter, they found that the typical desk location at Saunders Hall was receiving 120-foot candles from an eight-bulb fixture and 90-foot candles from a six-bulb fixture, proving that the rooms were indeed over-lit (Wolfe 2008, 1). They had succeeded in truly measuring what "too bright" meant, and they knew exactly what they wanted to do: reduce Saunders Hall lighting to healthy levels. After testing various scenarios, they determined that removing half of the lamps from every fixture would produce the optimum amount of light.

STEP 6: THE POWER OF THE PILOT: Since there was still a false perception that the changes might cause greater sacrifice, the HUB decided to perform a pilot implementation of the proposed project. Pilot projects address risk aversion, since any undesirable effects are small in cost, scale, and time frame and are easily corrected for the next iteration. Useful pilots must also be sufficiently substantial to reveal the benefits and pitfalls of wide-scale implementation, and they must be performed with community participation to generate feedback. This also allows measurement beyond economic and environmental factors and ensures the community benefits, creating an increased triple bottom line. More community members involved also meant that more people could advocate for full implementation of the project after the pilot was complete.

In soliciting participants, the department chairs requested their faculty and staff members to participate in the pilot that would take no more than twenty minutes of their time for delamping (removing lamps) and answering survey questions. With top-level approval, every person in the department was introduced to the concept of delamping by their department chair rather than by a student group. Within a few days they had over thirty volunteers, nearly five from each floor and department. This ensured that by the end of the pilot almost everyone in the building would have spoken with someone in his or her proximity who had his or her office delamped. Research shows that top-down requests from the department chairs may have been the single most important aspect of their efforts for successful wide-scale implementation.

McKenzie-Mohr (2006, 33) describes this phenomenon through the story of many farmers in the 1930s who suffered the loss of topsoil from their fields. The US government ran an educational campaign, with brochures outlining proven solutions like planting trees as wind screens to reduce the amount of soil being blown away. The entire first chapter of McKenzie-Mohr's book describes how simply giving people information, such as brochures, rarely works, and the campaign, like many others, failed to create change. When the government realized nothing was improving, they decided to go to a small farm within the region and physically help them implement the suggestions, creating a pilot that modeled the solutions for other farmers in their area. They hoped the farmers would talk with each other and become familiar enough with the new methods to implement them. This strategy accomplished what the brochures could not: when farmers saw the results, they adopted the solutions, and the new agricultural practices spread quickly.

Pilots and social modeling allow people the opportunity to get comfortable with the new processes at low risk, until they no longer seem new, and people actually overcome risk aversion and resistance to change. With many such successful case studies, it is not surprising that McKenzie-Mohr (2006) describes social modeling as one of the most powerful tools for encouraging widespread acceptance of a new concept, process, or technology. This was the case with the Sustainable Saunders delamping project. In addition to energy-use reduction that benefitted the economic and environmental bottom line, the pilot proved successful through a survey that showed that headaches, eye-strain, and stress decreased, resulting in an improvement in the social bottom line (fig. 10.1).

With an improvement in the triple bottom line of social, environmental, and economic benefit, the pilot was deemed successful. By the time the pilot was complete, all occupants knew what delamping meant and why it was beneficial to them and the university. They also knew that their leadership had approved the pilot, and since everyone had been invited to participate, there was a sense of inclusion.

STEP 7: WIDE-SCALE IMPLEMENTATION: Once the pilot was deemed successful, the community embraced delamping. The HUB had charted a course that accomplished a widespread energy-saving implementation without pushback in Saunders Hall.

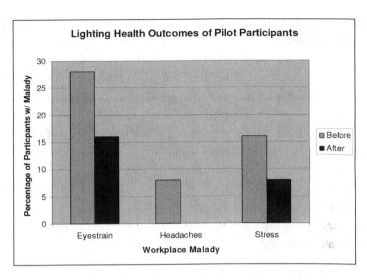

FIGURE 10.1. Lighting health outcomes of pilot participants (Wolfe 2008, 2).

STEP 8: IMPROVING THERMAL COMFORT WITH AIR CONDITIONING: The next big task was air conditioning. The HUB first spoke with the facilities engineers to determine what had been attempted before and what was believed to be ideal. They discovered that the air conditioning units were designed for the worst-case scenarios: hot days in the middle of the afternoon with rooms full of people. With a very limited number of engineers to run the entire campus due to ever-decreasing facilities budgets, they did not have the time or the means to measure variability throughout the day and had received complaints when they had tried turning the air conditioning off on previous occasions. The engineers were very supportive of attempting to reduce the air conditioning on an incremental basis, if they could liaison with the community to avoid the pushback. Wireless sensors called data loggers that read the temperature and humidity every minute in various rooms were placed throughout the building to monitor how the temperature changed throughout the day. The HUB identified that when the air conditioning was off, the building stayed cool for the morning, evening, and nighttime and was much warmer during the daytime hours. When the air conditioning was on, the temperature was hitting the engineers' goal of around 72°F midday, but it was much colder in the morning, evening, and nighttime. They found that timed air conditioning shutdowns were the answer.

The HUB originally tried to ensure temperatures did not go below 72°F until someone questioned why it was the ideal temperature. They researched the origination of the standard and were surprised to learn that the standard air conditioning levels corresponded to the amount of clothing people wore in offices, which is quantified by a unit of measure called CLO. For example, someone wearing a business suit has a CLO of 1 and prefers 72°F, which is why most air conditioning is set at that temperature. Interestingly, women would score a lower CLO unit if they wore skirts. The standard is set for men wearing suits. However, in Hawai'i most people wear short-sleeved shirts, shorts, and sandals, which are quantified by a CLO of 0.5 and should be most comfortable at 76°F. The HUB confirmed this theory at Saunders Hall by taking temperature readings throughout the building. A third of the occupants could control their own air conditioners, and measurements showed that they kept their rooms at an average of 76°F.

Once they had a goal in place, a pilot was performed. They decided that the incremental changes would begin in the nighttime hours from midnight to 5 a.m. with this window increasing slowly, gauging community acceptance at each level. Had they simply turned off the air conditioning at the level they thought optimum, with no intervals, there might have been pushback.

SMALL EFFORTS LEAD TO BIG CHANGE

The Sustainable Saunders Initiative was able to bring occupancy comfort into healthy levels and reduce the building's energy use by 24 percent, saving $150,000 every year for UH—all without spending any money. Their measurable success spread across campus as they presented their methodology and results at events, conferences, business meetings, schools, and organizations all over Hawaii. Everyone was looking for ways to save money during tough economic times in a state with the highest electricity rates in the country. The HUB became front-page news. They were no longer a small student group rallying for positive change—they were credible implementers with a valuable process that could put money into home and business owner's pockets across the state.

REPLICATING SUCCESS

Once campus had the proof of concept needed to implement delamping and air conditioning shutdowns on a widespread scale, the Sea Grant Program and the Center for Smart Building and Community Design created a lighting audit program, with student interns hired to begin performing lighting audits across campus. Facilities began asking buildings to volunteer for "Manoa Green Days," which meant that on designated holidays, weekends, and evenings participating buildings would have their air conditioning turned off.

They were overwhelmed at the supportive response from the campus and inspired when other organizations began reaching out to collaborate. They joined forces with the Architecture Sustainability Lab to do an energy audit of their largest campus building, Hamilton Library, and identified $700,000 in similar no-cost savings. They also worked with Engineers without Borders to design a water catchment system, Net Impact from the business school to host panel discussions, and the Geology Club grounds and landscaping staff to hold lunch party recycling drives.

The HUB continued to create new chapters by making weekly visits to other campuses in the UH system. Since they had expanded beyond Saunders Hall into the rest of the UH system, they needed a broader name, and so they renamed their group Sustainable UH. Their new mission statement was to become a world leader in sustainability education, research, and practices. They introduced a new goal for the next UH president's efforts on campus that followed their mission: to cultivate the UH system as a world leader in sustainability education, research, and practices. While there had been only seven presidential goals for years, the board of regents unanimously passed this eighth goal. Finally, the change they desired became valued. This initiative integrated sustainability into the broad, existing structure of the campus. A bridge between the grassroots organizations and the upper administration was formed. Beyond campus, there was greater alignment from top-level leadership and the growing efforts throughout the system.

INTEGRATING LESSONS INTO CURRICULUM

As new cohorts of students joined the HUB, they found that while they had a basic understanding of many of the same concepts, they used different words to describe the same thing, so they were spending valuable time translating to get everyone on the same page. They needed curriculum that would be an entry point to the information that would be accessible and interesting to any level of experience or interest, and that would provide the common language and quick decision criteria. In a HUB brainstorming session, they named the course Sustainability 101: Basic Training for Sustainable Action. The content included case studies from UH and Hawai'i, as well as national and global trends. The classes started with the simplest concepts and layered the language learned in early modules into later modules, building context for students. The course, rather than using a large document, was based on a PowerPoint presentation, with the main points included in the notes section so it could be used easily by all faculty members. The course modules were also developed so each could be taught as a stand-alone course, or the modules could be spread out and/or integrated into existing curriculum.

An online version of the course has been taught at Kapi'olani Community College, as a political science capstone internship class, in the Hawai'i Department of Education Construction Academy, and as a three-hour workshop through the Outreach College at UHM, which is a noncredit branch of the university. People from businesses, nonprofits, public and private schools, government agencies, the electrical and water utilities, and many other diverse groups have attended over the years. Those ranging from newcomers to experts gained knowledge from the material, as well as the opportunity to discuss real challenges with each other.

The future vision for this curriculum is to engage faculty to incorporate these modules into all streams of academia, creating Sustainability Across the Curriculum (SAC). SAC is modeled after the Writing Across the Curriculum (WAC) movement that swept through higher education across the country. WAC incorporates writing into all disciplines, rather than requiring only students in writing-focused degrees to be proficient in writing. In the same spirit, the idea is that no matter what paths students might take through higher education, they would graduate understanding the seven guiding principles. They wouldn't necessarily associate them with sustainability but, rather, would understand them as useful principles in the context of their own discipline. While this has only begun by tailoring the modules for small business, construction, and graphic design, the versatility is already proven, and the potential for the integration of these concepts throughout higher education is an exciting, and very possible, preferred future.

CORE PRINCIPLES OF SUSTAINABILITY EDUCATION MOVEMENTS

PRINCIPLE 1: THE TRIPLE BOTTOM LINE: Rather than considering people, planet, and profits as trade-offs, sustainability promotes simultaneous benefits to all three.

PRINCIPLE 2: CLOSED-LOOP CYCLES: Since all of nature operates efficiently in interconnected cycles, this module proposes that students should strive for cyclic human processes, such as turning waste into food or fuel. The HUB also applies this philosophy to human interactions: the goal is to turn negativity and conflict into compost for common ground. Hence, this module introduces the language of "close the loop" and "make it cyclic." Global examples include ecoindustrial parks for circular economies in China. Re-Use Hawai'i is a nonprofit that takes construction materials that would usually end up in landfills and resells them. Hawai'i also has a new law that electronics must be recycled by the manufacturer. Participants are then encouraged to think how they could close the loop on linear processes or relationships in their homes and workplaces.

PRINCIPLE 3: ECOLOGICAL/CARBON FOOTPRINT: Footprints describe the resources required by lifestyle choices. Methods and websites for calculating footprints for individuals, organizations, and businesses are shared, including discussions of metrics' strengths and limitations. Footprints of developed and developing nations are compared, and classes discuss what contributes to large footprints and what actions minimize them. Carbon taxes and cap-and-trade are introduced as policies that would provide incentives for widespread, large-scale footprint reduction. Case studies include Evolution Sage, a Hawai'i-based nonprofit that provides local carbon offsetting; Hagadone Printing, Hawai'i's only printing company that offsets all of its carbon production; and the greenhouse gas inventory at UHM, as well as many Sustainable Saunders projects.

PRINCIPLE 4: ZERO WASTE: This module introduces the concept of producing no waste by recycling, composting, repurposing, and reducing. Toronto, with its curbside compost pickup and unsorted curbside recycling, is used as an example of "closing the loop" for zero waste. Humboldt County, California, also provides enough energy from their municipal compost facility to fuel its entire wastewater treatment plant. Sustainable Saunders efforts for a plastic-free campus include dumpster dive waste audits, implementing biocompostables, and trayless cafeterias to reduce plate waste and water use.

PRINCIPLE 5: NET-ZERO ENERGY: Net-zero energy is the concept that buildings and campuses produce at least as much energy as they consume. This module demonstrates the difference between conservation and efficiency. Students discuss the role of renewable energy, and basic policies and financial models that make its implementation viable. Peak energy shaving and the role of electric cars are explored. The Sustainable Saunders energy audit, and the survey for Saunders Hall and Hamilton Library (including installing compact fluorescent lights, delamping, air conditioning shutdowns, and measuring occupant health and happiness) are all included as examples.

PRINCIPLE 6: LOCAL FIRST: "Local first" is introduced in terms of food miles and multiplier effects for healthy local economies. Benefits of "local first" thinking include smaller carbon footprints, healthier food, and increased community resilience. Global trends towards "local first" are described, including the case study of a local currency in Salt Spring Island, Canada. Campus case studies include building relationships with local vendors and farmers for food and services.

PRINCIPLE 7: CARRYING CAPACITY AND SELF-SUFFICIENCY: These terms are defined as the ability to be self-sufficient as a region, which has magnified importance for Hawai'i as the most remote population center on Earth. Hawai'i's dependence on imported fossil fuel, food, services, and goods is thoroughly discussed. The *ahupua'a* system was reintroduced as a unique model that inspired many members of the HUB and people throughout Hawai'i. Ancient Hawaiians divided the islands roughly into pie shaped wedges, each of which ran from the mountain to the sea. Each wedge, or *ahupua'a,* was self-governed and self-sustaining and thus honored its carrying capacity with amazingly intricate methods of cultivating land and aquaculture. As such, a population similar in size to the current Hawaiian populace could live completely self-sufficiently, with resources at every elevation within each district (see chapters 1, 3–5 of this volume for details of the *ahupua'a* system). It is a model that inspires us in Hawai'i but can also serve as a great inspiration for the world.

CONCLUSION

While the HUB was just one small group of students, a new project was born through their collaboration that could never have been accomplished alone. It opened new doors, and the synergy of different perspectives, voices, interests, and skill sets turned individuals and groups into networks. Their proof-of-concept projects earned them the reputation of turning ideas into action, and today their projects and partnerships have resulted in smooth implementations as the campus culture continues to evolve and embrace change. By sharing power, by recognizing it in everyone, and by giving it away, they gained it. The excitement of the process, the rush of creating something meaningful and new, and the fun spirit in which they worked and played together turned their networks into a true sustainability movement for higher education.

NOTE

1. It takes more than goals to actually create systemic change. Action must come from all parts of the system at the community or grassroots level, yet grassroots efforts that are not well linked to the institutionalized procedures and processes often result in groups failing to implement their goals (Innes 1992, 451). Thus, systemic change has a greater chance of success if both top-level priorities and mobilized grassroots community action are in alignment.

REFERENCES

College Student Educators International. n.d. "Change Agent Abilities Required to Help Create a Sustainable Future." http://www.myacpa.org/task-force/sustainability/docs/Change_Agent_Skills.pdf.

Coskun, M. 2008. *The Characteristics of Change Agents in the Context of Organizational Development*. Gothenburg: IT-University of Gothenburg.

Csikszentmihalyi, M. 1990. *Flow: The Psychology of Optimal Experience*. New York: Harper and Row.

Innes, J. 1992. "Group Process and Social Construction of Growth Management." *Journal of the American Planning Association* 58 (Autumn): 440–451.

Kelly, L., B. Cashore, S. Bernstein, and G. Auld. 2009. "Paying It Forward: Path Dependency, Progressive Incrementalism, and the 'Super Wicked' Problem of Global Climate Change." Paper presented at Climate Change: Global Risks, Challenges and Decisions, Copenhagen, Denmark, March 10–12.

Klein, J. 1990. *Interdisciplinarity: History, Theory, and Practice*. Detroit: Wayne State University Press.

McKenzie-Mohr, D. 2006. *Fostering Sustainable Behavior: An Introduction to Community-Based Social Marketing*. Gabriola Island, BC, Canada: New Society Publishers.

Princeton Review. 2010. "Princeton Review's 2011 'College Hopes and Worries Survey.'" In *Best 373 Colleges: 2011 Edition*. Princeton, NJ: Random House.

Rowe, D. 2002. "Environmental Literacy and Sustainability as Core Requirements: Success Stories and Models." In *Teaching Sustainability at Universities*, edited by P. Lang, 79–104. Frankfurt: Peter Lang International Academic Publishers.

Sandercock, L. 1998. *Towards Cosmopolis: Planning for Multicultural Cities*. London: Wiley.

Tierney, W. G. 2001. "The Autonomy of Knowledge and the Decline of the Subject: Postmodernism and the Reformulation of the University." *Higher Education* 41 (4): 353–372.

Wolfe, M. 2008. *Workplace De-Lamping*. Policy Brief 004, 1–2. Honolulu: University of Hawai'i College of Social Sciences Public Policy Center.

ELEVEN

It Takes a Village

Reflections on Building an Island School Garden

SUSAN WYCHE AND KIRK SURRY

On any given day at Kihei Elementary School, there is something special happening in the heart of the campus. This school, located on the island of Maui,
Hawaiʻi, is home to a thriving garden filled with Polynesian canoe plants and
a variety of European, Filipino, Latin American, and Native American fruits
and vegetables—all planted and cared for by more than 700 students, teachers, parents, and volunteers.

Working side by side with their students in this living classroom, teachers
translate science, math, social studies, and language arts into hands-on lessons.
During recess, students stroll through the strawberry patch in search of the
sweetest berries, or water their favorite plants with "rainmakers" made from
recycled empty yogurt containers. Their enthusiasm for pulling weeds, play-
ing pest detective, and maintaining compost bins is rewarded with fresh snacks
from the day's harvest and extra produce and seeds sent home to share with
their families. A parent stops by after school to chat with garden volunteers,
and takes home some of the herbs that need to be pruned back. A neighbor
drops off a couple of shovels that he salvaged and repaired the handles, thinking that the kids could use a few more.

The Kihei Elementary School Garden has created opportunities for intersections between teachers, students, and community that did not exist before

in Kīhei, and the impact extends beyond the simple activity of growing and eating one's own food. This may not be a "community" garden as the term is typically used, but that is the role that it fills in this neighborhood. It is a garden tended by people of all ages, sharing knowledge and resources and cultivating an intimate connection to nature, sustainable food sources, and better nutrition.

REVIVING A MOVEMENT ISLAND STYLE

When Alice Waters, the well-known owner and chef of Chez Panisse in Berkeley, California, started the Edible Schoolyard Project in 1995, she kicked off a national movement of school gardens, mainstreaming a growing national interest in organic foods, personal health, and sustainable living. Her one-acre garden and kitchen classroom became a model for thousands of gardens that have been installed since then, including gardens across Hawai'i and other Pacific islands (Waters 2008). School gardens, such as the one in South Kīhei, may have blossomed recently because of Waters' inspiration, but the concept of school gardens has a long history. Botanical gardens appeared in European universities at least as early as the beginning of the sixteenth century, and the one at the University of Padua, installed in 1545, continues to this day (Greene 1910; Miller 1904). These gardens included medicinal herbs as well as food, because gardens also served as the pharmacies of the day. School gardens spread throughout Europe, as other countries imitated the garden movements of Italy. In the United States, school gardens appeared in the late nineteenth century, beginning in 1869 in Boston. Since then, interest in school gardens has waxed and waned through the decades.

School gardens became a sign of patriotism during World Wars I and II, when growing food for local consumption allowed large farms to concentrate on crops for troops abroad. During times of peace, these gardens disappeared from all but experiential-based school programs that philosophically taught the "whole child," such as those influenced by Maria Montessori and Rudolf Steiner. It is not surprising that Alice Waters studied Montessori's approach and even taught briefly at a Montessori school before training as a chef. Nearly twenty-five years later, when she adopted a nearby middle school in Berkeley and worked with a small group of teachers and volunteers to plant an "edible schoolyard," she cross-fertilized the public school curriculum with both Montessori methods and her culinary training. That garden, and the attention she garnered for it, spurred a movement that continues today, fueled by the growing interest in the environment and sustainable living.

In Hawai'i and other Pacific islands, the school garden movement intersects with a traditional culture of sustainable cultivation that still remains. Where families have access to land, children learn to garden with their extended family in small family farms, backyard gardens, even patches along road right-of-ways. Coconut palm, papaya, banana, guava, citrus, and mango trees are incorporated into the ornamental plantings typical of older housing developments, and families still grow at least some of their own food.

This is the last remnant of what was once a thriving island culture. Today in Hawai'i over 90 percent of food and energy is imported. In the last few decades, as the cost of living has increased and jobs shifted from agriculture to service-related industries, more families have moved to apartments, condominiums, or houses with small urban yards, and the opportunity to garden has dwindled (La Croix 2001). For these families, school gardens may be the only opportunity the children have to learn about growing one's own food, lessons that would have been handed down formerly by parents and grandparents (Page, Bony, and Schewel 2007). Without the continuity of a family garden, the connection can be lost in a single generation.

School gardens not only help reconnect children with the land but also increase their understanding of the natural world, enhance their learning with hands-on activities, engage them with eating better foods, and create the link between young and old generations that urban living may sever. In a world of dramatic climate change, dependence on imported fuels, and international terrorism, school gardens also offer an opportunity to build safer, more self-reliant, and sustainable communities. These issues become even more compelling in the context of island life.

GETTING STARTED: A CAUTIONARY TALE

The concept of a school garden is simple in the abstract. Find a plot of land, add water, seeds, and compost, get kids planting and enjoying the harvest, and a successful school garden project is under way. But the reality is more complex and the undertaking more challenging than most people anticipate. This chapter tells the story of one of Hawai'i's most successful school garden projects and what was required to start and maintain a large school garden. The cost in time and money was far greater than most garden founders imagined. The impact has also been far greater, and that is what keeps any garden project growing. This is told both as a cautionary tale, to help spare others some of the lessons that were learned the hard way, and as a celebratory tale, to help other schools undertake what may be the most important change they can make to island school environments and curricula.

PLANTING SEEDS

Kihei Elementary School Garden started when a group of volunteers interested in sustainability decided that they wanted to address food security as one of the key areas for improving sustainability in their neighborhood. The group was originally focused not on school gardens but on a broad range of gardening projects—home gardens, neighborhood (*hui*) gardens, community gardens, and school gardens. This group held several educational meetings on the subject of gardening, led a backyard garden tour, started one *hui* garden but didn't know how to organize it, and searched unsuccessfully for a community lot. Though some of these efforts were successful, the effects were short-lived. The one exception was the school garden.

The project began simply enough, as these projects usually do. A member of the volunteer group was a substitute teacher at the school. She talked to the permanent second-grade teacher about setting up a small garden plot for her class. The teacher was enthusiastic, so a small group of volunteers showed up over two weekends to build three wooden beds, complete with soil, and a simple irrigation system. They used money out of their own pockets, plus some donations from local gardening stores and lumberyards. The school principal had indicated a back area of the lot, near one of the buildings, where the beds could go.

Once the garden structures were built, the group and the teacher helped students plant the beds, and the class returned a couple of times each month to monitor the progress of their plants, weed a little, and harvest some of the beans, tomatoes, and greens that survived. Everyone was pleased with the effort, other teachers took note, and aside from a few pests that showed up, some watering issues, and much of the harvest lost to inattentiveness, the project was deemed a success.

However, one problem was not discovered until after the beds were built and the garden was halfway grown. The school used automatic sprinklers that were set up to run on reclaimed water. Though the raised beds were irrigated with potable water, the larger sprinkler system oversprayed the beds at night when no one was around. When the group discovered this, they realized that

if the garden were going to be used to produce edible food for the children—
and not be just a science project—it would have to be moved.

A SEED TAKES ROOT

In the meantime, watching the excitement of the first group of second grad-
ers, other teachers expressed interest in participating. The principal was supportive and indicated a central area of the school where the volunteers could
establish a new garden. The volunteers decided to move the three wooden beds
to the new area and add a few more beds for other teachers. The idea was manageable and would have required very little extra effort to develop from the
original garden. But this is where a small, manageable garden project took an
unexpected turn.

To know how many plots to develop, the volunteers surveyed the school's
teachers to see how many were interested. To their surprise, over twenty teachers said that they would like to participate in the expanded garden. Even dou-
bling the original garden would not give everyone an opportunity to plant and
harvest, so the group recalculated and determined that the garden would need
to be four or five times its original size. One volunteer happened to be a professional landscape designer, so she laid out a new garden with enough new
beds to accommodate all the teachers who wanted to participate. The design,
built to serve 500 students, covered nearly 10,000 square feet, the full length
of the school's largest wing. Plans were drawn, and everyone was enthusiastic
to begin work.

As the project shifted from vision to reality, the group began to encounter
the challenge of scale. It was one thing to bring in two or three pickup loads of
compost and mulch for three boxed beds, but another to spread twenty cubic
yards of mulch, and to pay for three large truckloads. It was one thing to organize a small work party of four or five people for a couple of hours, and quite
another to recruit twenty-five to thirty people over several weekends. Having
a conceptual design in hand and some numbers about who would use the garden, the landscape designer wrote a $6,000 grant proposal to the local county
government, which was funded. But though that might sound like a significant amount of money for a garden, it barely paid for basic soil and mulching
materials, an irrigation system, trucking, tools, and the office supplies the project entailed. The expanded garden quickly became a major consumer of time
and money.

Installing a garden of this size also required the use of power equipment, so the volunteers and a few of the teachers built the garden when the school was closed, primarily on weekends and holidays. For weeks the children and their teachers skirted around what became a large construction project in the center of their school. But the waiting and watching only generated more interest.

When the garden was finally ready for planting, volunteers faced their most complex challenge. The number of teachers interested in participating had grown during the construction phase to twenty-eight teachers, representing approximately 650 students. Whereas the year before local businesses were happy to donate a couple of flats of vegetable starts, the same was not true when twenty-five flats were needed. Even using seeds and small starts would cost several hundred dollars, especially because organizers were committed to using as many organic heirloom seeds as possible. To make funds stretch further, volunteers asked for more donations of equipment from the community and were relieved when the community stepped up to meet the need.

Throughout the planning, logistics proved daunting. It would take two to three weeks of multiple planting sessions to accommodate all twenty-eight teachers, and each class would need several adults to monitor twenty-five to thirty students with trowels. If two classes were scheduled in the garden at the same time, then twice as many tools were needed. If fewer tools were used, and students rotated through tasks, more people were needed to monitor activities. No matter how planting sessions were organized, covering the number of classes with adequate numbers of volunteers seemed nearly impossible.

FATIGUE IN THE WEEDS

By the time planting was complete, the group had exhausted its core members and a dozen additional volunteers recruited for the project. Still, everyone was pleased with the results. Teachers commented on how well behaved and engaged their students were in the garden. And much to everyone's surprise, the students from the second-grade class that had worked in the original garden the year before remembered their experience and set to work like old hands. Planning and building the expanded garden had been a lot of work, but the volunteers felt it was well worth the effort.

The satisfaction of planting the garden, however, was short-lived. Volunteers had not anticipated how much maintenance the garden would require once planted, and they found that within a week they had to get back to work. The soil drained too quickly and needed daily hand watering for the tender sprouts. The automatic recycled water sprinkler system that had been turned off during installation was turned back on a few weeks later by groundskeepers. The beautifully mulched plot filled with Bermuda grass and other noxious weeds, and new plants suffered from mold and fungal infestations. Soon, sweet potatoes and other vines settled in and began to grow in earnest. Beans grew faster than they could be picked, and the garden required extensive fertilizing. The only thing that didn't threaten the garden that first year was an epidemic of pests (they came the next year).

CULTIVATE THE TEACHERS

Despite these challenges, garden maintenance problems were relatively easy to identify and solve. The problem volunteers were not prepared for, given the teachers' initial enthusiasm, was that few teachers brought their classes back after the garden was planted. Week after week, the garden sat unused.

The volunteers had imagined that, once planted, the garden would be used for a variety of activities and had arranged for teachers to call a volunteer coordinator if they needed assistance. Monthly garden meetings were also scheduled for interested teachers and parents. But weeks went by and not a single teacher called for assistance in the garden, and after just a few meetings, most teachers were too busy to get involved. Perhaps, the volunteers thought, rather than come when asked, it would be easier for teachers if volunteers posted regular hours. So they set up regular hours, but still no one came. The garden grew, the volunteers weeded and maintained as best they could, but classes didn't return.

To engage the teachers with what was happening in the garden, the volunteers started up a newsletter with news of what was growing, volunteer hours, and stories from the garden. After several frustrating weeks, when teachers expressed appreciation for the newsletter but still failed to come to the garden with their students, the volunteers finally asked them directly why they weren't coming. And from the answers the teachers gave, the volunteers realized that they had failed to address the biggest challenge of all: cultivating the teachers.

To the volunteers—all garden enthusiasts—it was obvious that the garden could be used in multiple ways to teach a variety of lessons in math, reading, writing, and science. But this was not apparent to the teachers, most of whom had little to no experience with gardening. Teachers felt time spent in the garden was an "add-on" activity to their required standards and benchmarks. Yes, they were interested, but already hammered by a program of cutbacks, budget-mandated furloughs, and pressure from administration to boost standardized test scores, teachers were reluctant to take precious time out to spend in the garden. One teacher even confessed that being seen in the garden with her class might be a concern later if her students did not perform as well as expected on tests. More important, the teachers simply didn't have time to adapt their lesson plans to a school garden environment. It was not yet, in their minds, an outdoor classroom.

The first turning point came when volunteers created a sample lesson plan, using garden plants as the subjects for measuring exercises that could be adapted to each grade level. The lessons were included in the newsletter, and within a few days, three classes visited the garden. Still, this was hardly the level of response volunteers had hoped for, given the number of teachers that the garden was designed to handle. The idea of developing curriculum for all six grade levels—when that was not their training or interest—was overwhelming to volunteers who were already overburdened by daily garden maintenance with no students or teachers available to help.

THE POWER OF HARVEST

Late in the school year, badgered by students who wanted to return to the garden, one teacher finally asked if volunteers could assist with a harvest. Thrilled that a teacher had asked for an activity, one member of the volunteer team who also happened to be a chef decided to make it a memorable event. He set up a tent, brought in tubs of water for washing, salad spinners, spices, salad dressing, and his own cutting boards and knives. He sent students out in teams to harvest basil, salad greens, cherry tomatoes, and green beans. Other volunteers made salsa and condiments from the garden the night before to serve with the fresh-picked vegetables. The chef quizzed students on the variety of fresh vegetables harvested and invited them to participate in preparing salads. The kids harvested carefully, listened to the chef intently, and devoured the vegetables they had planted enthusiastically. Crucial to the events' success, volunteers spoke with students about the harvest as part of a plant's natural life cycle and nutrition of fresh fruits and vegetables—even cleanup activities became part of a class lesson on soil building and the roles of decomposers during the process, when students took their leftovers to the compost pile.

This was the major shift that the project needed. When the other kids saw the tent, the chef's uniform, and the fun their schoolmates were having, their teachers heard about it. When teachers realized that a harvest party could underscore lessons they were trying to teach in the classroom, they were willing to schedule time for the garden. Requests from all of the teachers who helped plant the garden came flooding in. For three weeks, volunteers guided student harvests and salad making while encouraging students to try a bite of everything they grew. Even the most reluctant children were willing to take at least one bite, and most were eager to munch on fresh vegetables and salad greens they grew themselves. The harvest parties hit the mark with students and teachers, and the project ended the year on a high note.

SUMMER BREAK: REST AND REFLECT

In Kīhei, as on most tropical islands, the intense heat of summer provides a natural time to allow a garden to rest. That summer, the volunteers rested. They let the weeds grow, and they talked about what they would do differently next year. More funds were needed as the original $6,000 grant was spent. They also knew that they needed someone who could work during the week, as most of the volunteers had full- or part-time jobs and could not keep up the effort that the first year required.

Everyone agreed that the work had been too much. But they also knew that they didn't want to shrink the garden or exclude any of the teachers who were interested. Instead, they wanted to find ways to get teachers more involved in the garden, even though it meant more work for volunteers, at least in the beginning.

At the same time, many members of the original volunteer group burned out. They wanted to participate in a greater variety of projects and felt that the school garden was taking too much time and energy. Just when the garden needed more help, the core group of volunteers contracted.

THE CHEF CONNECTION

Though the volunteers had rested, the garden had not. Weeds grew, pests moved in, and the watering system failed in the heat of summer. Tools had broken and needed repairs. Compost, fertilizer, and other supplies were used up and needed replacing. To get the garden ready for next year would require another round of funding. If it hadn't been clear the first time, the volunteers could see it now: it takes a village to run a school garden.

From the start, volunteers had heard from other school garden organizers that they would need a garden coordinator. That first year, however, the group was so busy covering the basic expenses of the garden itself that a garden co-ordinator seemed far beyond their reach. Even a part-time coordinator cost more than they could afford. But it was clear that if this project was going to survive long term, it needed someone to provide continuity to activities; someone dedicated to bringing together teachers, students, and volunteers; someone to build the village of support the garden required. To pay for this, the project needed a major benefactor, a long-term grant, or other resource to cover a coordinator's salary.

Though the group discussed this topic repeatedly, they came up with no solutions. They kicked around ideas but had little time to spend on develop-ing and implementing them. The solution came not from careful planning but from the enthusiasm with which the project's volunteer chef had spoken to other chefs with whom he worked. Two of his colleagues visited the garden to see for themselves what the commotion was all about. After just a few hours of digging in the soil and talking to students about the food being grown in the garden, they made a major commitment to provide monthly funding for a part-time coordinator.

"Everyone has a voice in this community, and this is one way that I choose to use my skills to give back and contribute to the island 'ohana," said chef Brian Etheridge, owner of Capische Restaurant. Chef Dan Fiske, a private chef and Brian Etheridge's partner as a sponsor for the garden, agreed: "The garden is the heart of the campus. It's amazing the way it makes you feel when you step foot inside!"

The two chefs helped prepare the garden during the next school year and then spent a week with several hundred students harvesting fresh basil, thyme, oregano, onions, and tomatoes from the Pizza Garden and demonstrating how to make caprese salad and a pizza sauce for homemade pizzas. While slicing pizzas for sampling, they quizzed students on addition and division problems. Though the late spring days were long and hot, the chefs vowed to return for future harvest parties. Their love of fresh food brought the garden project full circle and helped students and teachers recognize that growing the food is only part of the equation.

This connection makes sense, of course. As chefs, they were aware of the need for local food sources and a new generation of farmers and young chefs to grow food and prepare it. Their businesses also rely on having clients who appreciate fresh, quality food—which means that the more children are educated about good food and nutrition, the better customers they will be in the future. But neither of these reasons is what won the chefs' support. It was the enthusiasm, interest, and appreciation they received from the students when they came to the garden—the same thing that had kept the overworked group of volunteers showing up week after week. The chefs also saw that the garden needed a garden coordinator, and with their support, the Kihei Elementary School Garden moved into a new, more stable phase.

COORDINATING A MOVEMENT

The next year started with hiring the part-time garden coordinator. The new hire was a former volunteer with no formal training as a teacher or as a horticulturalist—a public relations specialist by training. He had been an enthusiastic garden volunteer during the first year of the school garden project, grew his own food at home, and became increasingly more knowledgeable about gardening as the project grew. To further develop his knowledge of gardening, he enrolled in the local university extension program's master gardener classes.

That third year even more teachers signed up for planting. But this time, the teachers and the students knew what to expect, as did the volunteers. By the time the garden coordinator settled in, more teachers were coming to the garden, and new recruits were showing up from both the community and the students' families. The garden coordinator used his communication skills to help recruit more volunteers, celebrate the garden's successes, and share with the broader community the incredible impact that gardens can have on student learning, eating habits, and interest in growing their own food. He kept the newsletter going with the help of the other volunteers and focused on connecting with students, teachers, parents, and staff.

Equally important, he began arranging teacher meetings to review pacing plans and benchmarks, helping teachers develop garden lesson plans that matched standards at each grade level. The results were obvious. Teachers who were previously hesitant to explore outdoor learning began regularly visiting the garden for a wide range of curriculum-specific lessons, including Native Hawaiian studies, the life cycle, nitrogen and water cycles, pollinators, decomposers, plant species and parts, and scientific investigation. And as the teachers realized the potential of an outdoor classroom to engage their students in the learning process, they found more ways to translate their studies into garden activities.

In evaluations, thank you letters, and casual comments, teachers recognized the impact of the program on students. "The garden has helped my students learn the parts of a plant, how plants grow, and how to take care of plants," wrote one teacher. Another commented, "It has supported our science standards and has given the children hands-on learning experiences that are *so* extremely valuable!" Teachers noticed other benefits as well. Some students who were quiet in the classroom became more social and engaged in the garden. Special education students often responded so well that teachers would send them into the garden with individual tutors. And as one teacher remarked, "They always seem to remember garden days the best."

LEARNING THROUGH CULTURAL DIVERSITY

Students and teachers were not the only ones who discovered new things in the garden. The volunteers who helped design and install the garden also learned. They had started the project with vegetables and herbs they knew best, primarily European species—green beans, lettuces, tomatoes, and herbs. But eventually, the influence of students, teachers, family members, and local experts expanded what was grown in the garden, including Asian greens, beans, herbs, and Polynesian "canoe plants" brought by Hawai'i's earliest settlers: bananas, papayas, *kalo* (taro), *'olena* (turmeric), and other culturally significant plants. These plants not only survived local tropical growing conditions better than their European counterparts but also provided opportunities to integrate the garden with what the students were learning in their social science, history, and cultural lessons and to reflect what many students ate at home.

The plants valued by local cultures also brought members of the community into the garden to share their techniques for cultivation and use. In the second year, a Native Hawaiian, who was a spouse of one of the teachers, demonstrated the tradition of harvesting *kalo* and pounding poi. Parents of students, school staff, and groundskeepers would stop by to admire the sweet potato plants or Asian winged beans—the seeds were a gift from one of the salesmen at the irrigation store—and share their secrets for growing or cooking them. Slowly the garden became a community site for shared learning.

TIPS FOR CREATING A SUSTAINABLE SCHOOL GARDEN

As the Kihei Elementary School Garden thrived in its third year, the volunteers began working with additional community groups to start gardens at other area schools. Having survived one garden startup, and with a second launch under way at a neighboring school, they acquired knowledge that would have been useful prior to starting this fulfilling yet challenging garden project. Some of the realizations come with garden experience—understanding that, once started, gardens are an ongoing responsibility, continually posing new challenges. Some of this knowledge is organizational and community experience—understanding that a significant part of the effort will be cultivating the people who are needed to make the garden project a success: students, teachers, parents, administrators, local officials, local garden and landscape vendors—working with a broader community that the garden both supports and depends upon. There was also much to learn about landscape installation, media relations, grant writing and administration, and ongoing funding.

The Kihei Elementary School Garden has been featured in local newspapers and national news media, had a local film made about it, and was even visited by the governor of Hawai'i, who now uses it as an example when discussing sustainability issues. Despite becoming a model program and providing great returns to those who started it, the journey has often been rocky and the problems challenging. For this reason, the volunteers have developed a few basic tips for starting school gardens, described below.

Do Your Homework

Before taking the leap into school gardening, visit other schools that have managed successful programs and check online for local school garden networks.[1] Whether visiting schools in-person or browsing through their websites, this research can be invaluable when planning the first moves, saving time and helping to manage obstacles along the way. Pay special attention to gardens that are in climate zones similar to your own. Books can offer more in-depth information about building school gardens and developing curricula to make them relevant and effective (for two good starting points, see Stone 2009; Bucklin-Sporer and Pringle 2010).

Create the Vision

Involve teachers, students, and parents from the beginning as you develop a vision for what the garden will be: how big, how many teachers and students will use it, and where it will be located. School gardens that are influenced by student-driven brainstorming sessions and sketches of how students envision their new edible schoolyard will develop buy-in from students and teachers before the first shovel of soil is turned. This is especially important for projects that are introduced from outside the school.

Research Curriculum

Ask teachers to review curriculum that they can teach in the garden, as well as specific fruits, vegetables, herbs, or other plants that they may relate to their lesson plans. At first, review pacing plans and standards to identify ideas. Reviewing these plans with several teachers from each grade level will develop a greater understanding of class-level learning objectives and help individual teachers adapt the vision and design elements to match their objectives.

Build Financial Resources

Identify adequate funding not only to build the garden but also to maintain it, and decide who will shoulder that responsibility—the teachers, the school, an external community group, the Parent Teacher Association (PTA), or some combination—perhaps a dedicated garden group with representatives from all of these. Consider partnering with local chefs and restaurants, landscapers, landscape supply companies, and others who can provide ongoing support, as well as professional knowledge.

Engage Volunteers

Be sure to collect names and contact information of the people and businesses that support your efforts and put them into a database that can be used for multiple purposes. Find lots of opportunities to say "thank you," and to keep in touch with volunteers. Generous supporters and volunteers are a school garden's most precious resource. Look for opportunities to reach out to them.

Select a Visible Location

The saying "out of sight, out of mind" certainly holds true for school gardens. Wherever possible, gardens should be central, not pushed to the edges of the school grounds. This consideration also helps with security. Gardens in the heart of a school are less likely to attract unwanted attention than those on the outskirts of school grounds. You may decide to select one location that is a busy, central spot on campus (i.e., adjacent to the cafeteria or administration office) or several locations that are visible and easily accessible to classrooms or specific grade levels on a larger campus. A centralized garden will be easier to supply, water, and maintain, but smaller pocket gardens may feel more manageable and proprietary for individual grades or classes.

Ask the Experts

Once you have a vision for what the teachers and students want in the garden, develop a garden design. Students may work with professional designers to prepare designs (depending on availability and funding), but always be sure that a professional reviews the final design before the garden is installed. A scaled drawing will prove invaluable for calculating mulch and soil, laying out beds, and other tasks where the volume of material needs to be accurate. Make sure there is adequate room for wheelbarrows, for students to cluster around beds and not trip over one another, and for truck access to bring in fresh compost, soil, and other supplies. Designs should provide access for people with disabilities. A beautiful design may also help you "sell" your vision to donors, grant sources, and other funders. Before installing the garden, show your design to local gardeners, farmers, and irrigation specialists to get recommendations on fruits and vegetables that grow well in your area and identify the necessary irrigation supplies and tools you will need to care for the crops. Before you dig, work closely with the school's maintenance managers and utility companies to mark utility, data, and irrigation lines so those areas can be avoided.

Gather Local Resources

Sustainable gardens will regularly need free or deeply discounted materials and supplies from resources in the surrounding area. Develop relationships with local gardening supply and hardware stores for tools and seeds, grocery stores for produce scraps, carpenters for used or surplus building materials, tree trimmers for mulch, and landscape supply stores for irrigation parts and soil amendments.

Plan Workdays Wisely

Garden preparation is hard work, and there are safety risks involved. Make sure that power equipment is kept away from volunteers, especially young children. Schedule major workdays when students aren't present (especially for younger students). Make sure you have liability forms, photograph release forms (for press releases so you can use any photos in outreach), shade shelters (if there are no trees or other shade nearby), water, fruit and other snacks, adequate tools (or ask volunteers to bring their own), first aid kit, camera, a table and chair to sign in volunteers, and adequate signage for nearby roads. Once the heavy job of earth moving, tree removal, and installation of main irrigation lines is completed, younger students can be involved. Developing a garden is a wonderful activity for bringing parents together with their children and teachers, and everyone benefits from the exercise.

Plan your workday activities carefully. Have a core team arrive early, and make sure that you have several team leaders who know the plan. An "Order of Operations" list is helpful if you are planning more than one major task, or starting a new garden. Have it posted at the check-in table and have a site manager assign volunteers to workstations or project teams based on skill levels.

Share Responsibility

One of the most difficult parts of growing a garden is remembering to make the work sustainable, too. Set up regular meetings to discuss garden maintenance, garden activities, funding, and volunteer recruitment. These are the mainstays of keeping your garden growing once it is built. Delegate these tasks to a broad group of individuals to avoid volunteer fatigue.

Share Harvests, Stories, Photos

Share the harvest. Share the food with students and their families, and with volunteers who come to work in the garden. There are always plants needing harvest, once a garden gets started. Gifts of food bring enormous goodwill and prevent hard work from going to waste.

Share the stories and images of the children, teachers, and volunteers in the garden—these personal reflections will be some of the most precious treasures cultivated in the garden. The project will be richer for sharing those moments with the broader community. Photos of students in the garden will be greatly appreciated by all those who contributed to the garden along the way. Blogs, websites, and newsletters are great avenues to share these moments, provided you have volunteers, teachers, or students who have the skills to develop and maintain them.

Build a Village

Last but not least, make sure that the garden project has the support and commitment of school leaders, the PTA, teachers, and other community groups before beginning. There will be plenty of work for all, and success depends on all of these groups working together. Gathering support early on also helps invest these groups in the success of the garden, rather than placing the responsibility of the garden into the hands of one group. This network will be invaluable as the workload grows.

As the story of Kihei Elementary School Garden illustrates, jumping in too quickly can start a chain reaction of challenging issues. The most important advice volunteers with the project offer is to start small and manage each expansion mindfully. By growing the garden and its village in manageable phases, one builds a solid, well-staffed garden that can survive the vicissitudes of pests, weather, and people who visit it.

SPECIAL CONSIDERATIONS FOR ISLAND-BASED SCHOOL GARDENS

As the newer participants of the garden project quickly learned, gardening on the mainland is different than on an island, where many problems are unique to specific island environments. Therefore, it is important to address island-bound challenges for building and maintaining gardens.

Soil Issues

Gardening on an island can be challenging and expensive if you are relying on store-bought soil mixes and amendments. On Maui, there's often a waiting list for large loads of soil and compost because commercial composting facilities are limited. Soil quality can also vary greatly from one side of the island to another, and contamination from downhill runoff is common.

Always test soil before planting, and then supplement with as much organic matter as can be generated onsite. Don't depend on what people from other garden sites tell you, as soil quality and chemistry can differ dramatically from one side of an island to another. University extension programs offer inexpensive but useful soil tests. Start compost piles and vermiculture bins to keep a constant supply of amendments on hand. Grow nitrogen-fixing cover crops and dynamic accumulators that help to attract micronutrients from the subsoil, using the plants that thrive best in your area, especially native plants if appropriate ones are available. If you start your garden in an area of a school that has been maintained using standard landscape practices, you may find that the soil is depleted of nutrients and will need extra amending.

Invasive Species

Islands often have fragile ecosystems that can be thrown off by seemingly innocuous plants. Some foreign or exotic plants propagate rapidly in Hawai'i's year-round growing environments, and their seeds or plant parts can travel long distances to naturally forested areas; these are the plants that are termed invasive and they can be costly to control. If uncertain about the status of a new species being introduced to the garden, check with local university extension programs or the regional department of land and natural resources. Some islands also have invasive species committees and much useful information is available online.[2]

Material Cost

Everything is expensive on an island: mulch, building materials, equipment, and tools. Seek local resources, surplus, and reclaimed supplies wherever possible to save money and teach students about sustainable, reusable resources.

Seed Saving

Food security is a real issue on islands. Teach students about seed saving and develop a system for categorizing and storing garden seeds throughout the growing season.[3] Over time, these seeds will represent plants that grow well locally. Seed saving solidifies sustainability and brings students closer to their food sources as they trace lineages and preserve varieties of their favorite fruits and vegetables.

Cultural Awareness

Most public schools have an abundance of traditional garden knowledge and experience among students and their families that school gardens can access. Islands have long-standing histories of migrant workers who travelled from other islands with similar climates who have brought favorite plants to their new home. Including plants from different cultures provides opportunities to focus lessons around culture, language, geography, and local environment. Garden spaces can honor different cultures, cuisines, and "useful" plants, and make history lessons come alive.

Food Safety

Everything grows well in most island environments, even pathogens. Learn about good local agricultural practices and research information about food-borne diseases, island-specific pathogens, and how to keep them out of the school garden. Again, the local university extension program is usually the best first stop for help with diagnosing issues and pests and finding effective treatments (Hollyer et al. 2011). Local garden clubs, sustainability groups, and most of all local farmers will be able to provide helpful information and wise advice.

Water Efficiency

Water resources on islands are often scarce and expensive, especially during a time of increasing droughts and climate change. Consider using drip irrigation systems that are set with timers to water during early morning and late afternoons, when less water will be lost to evaporation. Teaching students about water conservation can be a lesson that they can take home to their families. Some schools are also experimenting with closed-loop systems, such as aquaponics, which is an excellent addition for demonstrating maximum efficiency with water and plant fertilizer resources, while producing vibrant gardens grown in an organic, soil-free medium.[4]

GROWING OUR FUTURE

Those volunteers who started the Kihei Elementary School Garden—a substitute teacher, a former project manager with a large software company, a landscape designer, a chef, and a public relations expert—all had their moments of exhaustion and burnout.

But they saw disengaged students come alive in the garden, watched others be transformed by their experience tasting their first fresh-picked bean or tomato, and observed dozens of students digging happily in garden beds looking for earthworms. After just the first year, they knew that school gardens have an amazing capacity to transform education.

They also saw how a school garden gives students a connection to healthy food and an understanding of how that food is grown. It gives them a connection to the land, to the cycles of life, and to something that will continue to shape their views of nutrition, health, and the environment. Most important, a garden makes learning about these things engaging in a way that cannot be accomplished by books or computers. For these volunteers, the question is no longer whether schools should have school gardens—because the answer is obviously "yes"—but how soon can the village get started?

NOTES

1. For Hawai'i, see the Hawai'i Islands School Garden Network (http://kohalacenter.org/hisgn). The Collective School Garden Network (http://www.csgn.org/index.php) also carries extensive information that will be useful for new school garden projects. One of the best sites, which deals with school gardens and broader issues of sustainability, is the Center for Ecoliteracy (http://www.ecoliteracy.org/).

2. An excellent resource for the Hawaiian Islands, with links to others for the Pacific region, is the Hawaii Invasive Species website (http://www.hawaiiinvasivespecies.org/).

3. Books and articles on seed saving are available in libraries and online. The World Vegetable Center offers a short but useful manual that focuses on more tropical vegetables (http://203.64.245.61/web_docs/manuals/save-your-own-veg-seed.pdf).

4. Aquaponics has become very popular in Australia, and works well in island environments; the Aquaponics.net website (http://www.aquaponics.net.au/education1.html) provides a useful starting point.

RECOMMENDED GROWING GUIDES FOR TROPICAL ISLANDS

Creasy, R. 2000. *The Edible Asian Garden*. Boston: Periplus Editions.

Ellis, B. W., and F. M. Bradley, eds. 1996. *The Organic Gardener's Handbook of Natural Insect and Disease Control*. Emmaus, PA: Rodale Press.

James-Palmer, R. 1987. *Maui Organic Growing Guide*. Chelsea, MI: Bookcrafters.

Nugent, J., and J. Boniface. 2004. *Permaculture Plants*. White River Junction, VT: Chelsea Green.

Oshiro, K. 1999. *Growing Vegetables in Hawai'i: A How To Guide for the Gardener.* Honolulu, HI: Bess Press.

Toensmeier, E. 2007. *Perennial Vegetables: From Artichoke to "Zuiki" Taro, A Gardener's Guide to Over 100 Delicious, Easy-to-Grow Edibles.* White River Junction, VT: Chelsea Green.

REFERENCES

Bucklin-Sporer, A., and R. Pringle. 2010. *How to Grow a School Garden: A Complete Guide for Parents and Teachers.* Portland, OR: Timber Press.

Greene, M. L. 1910. *Among School Gardens.* New York: Russell Sage Foundation.

Hollyer, J., F. Brooks, L. Nakamura-Tengan, L. Castro, J. Grzebik, M. Sacarob, V. Troegner, D. Meyer, T. Radovich, L. Morgan Bernal, and D. Kishida. 2011. "Student and Food Safety: Best Practices for Hawai'i School Gardens." College of Tropical Agriculture and Human Resources, University of Hawai'i at Manoa. http://www.ctahr.hawaii.edu/oc/freepubs/pdf/FST-45.pdf.

La Croix, S. 2001. "Economic History of Hawai'i." http://eh.net/encyclopedia/article/lacroix.hawaii.history.

Miller, L. K. 1904. *Children's gardens.* New York: Appleton.

Page, C., L. Bony, and L. Schewel. 2007. *Island of Hawaii Whole System Project Phase I Report.* Boulder, CO: Rocky Mountain Institute. http://www.kohalacenter.org/pdf/hi_wsp_2.pdf.

Stone, M. K. 2009. *Smart by Nature: Schooling for Sustainability.* Healdsburg, CA: Watershed Media.

Waters, A. 2008. *The Edible Schoolyard: A Universal Idea.* San Francisco: Chronicle.

EPILOGUE

Living Like an Island

What the World Can Learn from Hawai'i

JENNIFER CHIRICO AND GREGORY S. FARLEY

Sustainability is frequently defined as the triple bottom line, meaning a focus on social and environmental as well as economic concerns. The triple bottom line is often illustrated with a Venn diagram, as shown in figure 12.1. The three aspects of the triple bottom line overlap in only one region, and that one region, interpreted as "true" sustainability, is very small relative to the sizes of the circles themselves. The Venn diagram shown in figure 12.1 also represents the inherent tensions within sustainability conversations and demonstrates the need for compromise with regard to one of the three circles if we are to move towards a more sustainable future.

However, the three circles on the Venn diagram do not need to overlap in this way to illustrate sustainability. When students are asked to think about the diagram and redraw it, many of them will draw three concentric circles, much like an archery target (see fig. 12.2). For many students, the outermost ring is labeled "environment," reflecting students' core understanding that the wealth of nations, and the societies that rise from economic principles, are built upon natural resources. Students still understand the value of compromise, but their compromises will often favor environmental causes and seek to reduce the influence of economics.

What Hawai'i shows us is a third way to arrange the rings that define sustainability. The three traditional rings again overlap, but a fourth outer

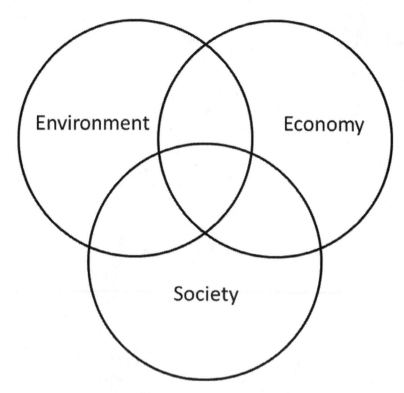

FIGURE 12.1. A traditional approach to sustainability: three overlapping circles representing the triple bottom line. The small area in which all three circles overlap is interpreted as "true" sustainability.

ring, culture, encompasses all other causes (see fig. 12.3). We refer to this as the quadruple bottom line, which also reflects a systems-based approach. The outer ring, culture, is well defined, whereas other authors may put other labels in that space (e.g., life, planet, environment, or happiness).

This is a lesson from Hawai'i: taken together, the case studies in this book highlight that Hawaiian efforts in food production, food security, water systems, and buildings are most successful when they are respectful of, and often rooted in, Hawaiian culture. The resurgence of native taro planters improves access to local food, makes better use of land and water resources, returns a critical component of nutrition to Hawaiian diets, and reduces the need for imported, and less nutritious, food. New homes, in order to be energy efficient and provide healthy structures, are now including practices that Hawaiians learned thousands of years ago, when buildings were designed to take full

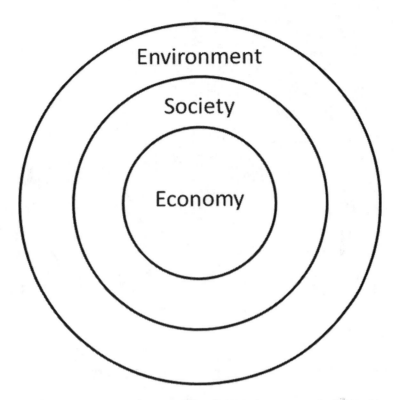

FIGURE 12.2. An alternate sustainability diagram: three concentric circles, with one core sustainability component (here, environment) encompassing the other two components of the triple bottom line. Note that it is also possible for the two inner rings to be of equal size or to overlap incompletely or not at all.

advantage of the prevailing winds, available sunlight, and local land and water resources.

This makes sense from an ecological perspective. Historically, Hawaiians developed processes that integrated the local island ecology into everyday living. They worked *with* the land rather than *against* it. Hawai'i, like all islands, has limited resources, and precolonial Hawaiians had to live within the limits imposed by nature to ensure their survival and long-term sustainability. Furthermore, unlike island nations with nearer neighbors, Hawai'i is so isolated that its people could not rely upon regular trade to make up for local ecological limitations. Instead, they had to live successfully within the parameters set by nature, and their cultural practices reflected those limits.

The success of precolonial Hawaiians, although not unique among island groups, does stand in stark contrast to the fates of other island populations

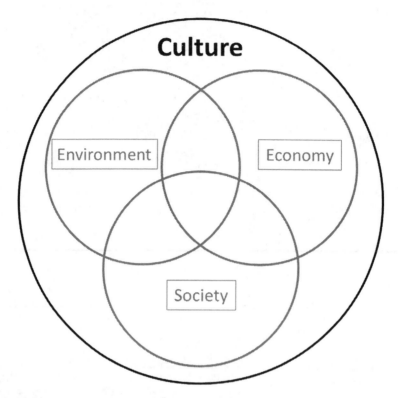

FIGURE 12.3. An emerging model of sustainability representing a quadruple bottom line: three overlapping circles, with culture encompassing all other components. Again, it is possible for inner rings to be arranged differently; the diagram represents the conclusion that sustainability is possible only when it draws strength from, and is respectful of, the culture seeking sustainability.

throughout the Pacific and around the globe. Nowhere is that contrast more stark than on Easter Island, where sustainability was achieved, and the population was eventually eliminated. Although new interpretations of the history of Easter Island (Hunt and Lipo 2011) are challenging traditional ones (e.g., Diamond 2005), all that differs in those interpretations is mechanism: the root cause of Easter Islanders' demise was most likely shrinking ecological limits to the point that islanders did not, or could not, live within them.

Although the basic human needs experienced by precolonial Hawaiians, such as the need for food, water, and shelter, have not changed over time, modern societies, including Hawai'i, have additional "needs" that are perceived by society as just as vital: electricity, running water, and easy access to food are the underpinnings of developed nations and are sought by developing nations. The meaning of sustainability today, therefore, includes those needs, and indeed, many definitions of sustainability implicitly or explicitly defer to development. The chapters in this book argue that technological approaches to sustaining large, modern populations must be respectful of local cultures and that the connections among the economy, environment, social issues, and culture must be fully integrated and addressed.

The integration of these components is described throughout the book. Water use, reclamation, natural processing, and reuse are vitally important to the economic success of present-day Hawai'i; better stewardship of water resources allows more natural water flow for Hawaiian planters, many of whom are forging a resurrection of the *ahupua'a* system, which was central to precolonial Hawaiian success. Improved stewardship of wastewater also improves the health of nearshore reef and fish populations, which, in addition to being a vital food resource for Hawaiian residents, also represents a cornerstone of the Hawaiian tourism industry. Even that industry, which brought a seemingly unchangeable and deeply unsustainable model to Hawai'i, is showing recognition that it, too, relies upon the natural beauty of Hawai'i and must therefore participate in sustainability efforts. Like native agriculture, Hawaiian tourism must become fitted to the land it draws from, or it will contribute to its own demise.

Energy production remains a critical question for isolated communities everywhere. In Hawai'i, as elsewhere, climate change poses a significant threat, and island groups must work to set an example for the rest of the world and to eliminate their own contributions, however small, to this global problem. Alternative, or "clean," energy will be vital to islands, and it will be important for island groups to capitalize upon the natural resources they have—again, similar to that of precolonial Hawaiians. Wind and solar energy production are already well established in Hawai'i and are growing as a proportion of the energy supply; additionally, efforts are being made to capitalize on the boundless energy of ocean waves.

The ubiquitous interconnectedness of Hawaiian culture that is woven into the fabric of the local island society poses a question: Are culture, people, environment, and economy inseparable? Elements of this book suggest that the separation of these concepts may, in fact, be false—that to be truly successful, and sustainable, a population must regard itself as part of the local environment. The Hawaiian creation myth (see endnote 2 in chapter 1) asserts that the first man was brother to the taro plant, implying a close connection between people and nature. The suggestion for isolated populations anywhere, but perhaps especially on ecologically limiting islands, is that modern humans need to relearn to see themselves as part of, rather than apart from, the natural world.

Sustainability in Hawai'i, then, offers three lessons to the rest of the world, and particularly to other island and isolated communities. First and foremost, the story of Hawaiian sustainability tells us that *we should look to history for practices that our forebears developed if we want to be sustainable.* This is, in some ways, intuitive: historically, preindustrial societies in all locations were more ecologically sensitive than modern societies are today. While modern society has made great strides in technology, which has resulted in high "growth" and seemingly better overall public health, the lack of ecological sensitivity over the last several hundred years has led to severe pollution in fresh water and all of the oceans, vast amounts of waste and thousands of landfills, inefficient buildings that have huge environmental impacts, depleted soils for growing food, and a strong dependence on fossil fuels and imported food.

Looking to history has additional benefits as well. Primarily, a historical approach to sustainability means that we need not reinvent the wheel. This is not to argue against progress, or against new, innovative ideas that cause economies to thrive—only merely to acknowledge that the breadth and depth of the human experience offer lessons that can be adapted to modern challenges. In addition, looking to history will help preserve culture and celebrate diversity. A principal lesson from ecology is that diverse ecological communities are more resilient to change, and more stable over time, which offers lessons for twenty-first-century humanity. Hawaiian sustainability looks forward, while also looking back and learning from a rich cultural heritage, and offers hope that other island groups, other isolated communities, other nations, and ultimately, the human population can create a more sustainable future.

The second lesson from Hawai'i is that *true sustainability requires reaffirming the place of humans in the natural world.* The growth of modern society has been fueled, in part, by an ever-increasing separation between humans and the environments they live in. This has had many benefits, particularly for developed nations. Examples are particularly rich in public health, which has improved, and infant mortality rates, which have declined. However, this increasing separation has also meant that humans often regard the environment as separate from themselves, and as a result, many people do not know where their food is produced or where their waste goes. This is problematic for long-term human survival. When societies fail to recognize the environments that support them, then they are at risk for hard lessons about food security, water access, and public health.

Tangential to this lesson is the idea that we must scale in tandem our understanding of economic growth and our understanding of ecological health. The late nineteenth and twentieth centuries were marked by rapid economic growth, and during the late twentieth century, particularly since World War II, the study and pursuit of growth in such markers as gross domestic product (GDP) as a measure of economic health came to overshadow ecological concerns. In short, the pursuit of profit topped the list of concerns in developed nations, to the detriment of the environment, and the developing world is currently following that model. What was lost during that era was a fundamental fact: GDP, and other measures like it, are really measures of the consumption of natural resources. We cannot continue to degrade natural capital at the current rate and still hope to sustain a global population in excess of 7 billion people. Only by recognizing and addressing the environmental and social impacts of growth can the economy truly sustain itself in the long-term and support a growing human population.[1]

The third lesson from Hawai'i is that *we must understand sustainability from a systems perspective.* Aldo Leopold urged us to "think like a mountain" or, as demonstrated by this book, "think like an island." Perhaps, instead, it should be to "*live* like an island"—to see the whole system, to recognize the importance of every part of the system, and to appreciate the interdependencies of the system. For other islands and remote communities around the globe, Hawai'i teaches us to recognize that systems have limits, that humans are a part of that system and not separate from the whole, and that every part of the system is interconnected and depends on the other parts for survival and long-term sustainability.

NOTE

1. Globally, multiple metrics attempt to assess systems-level sustainability and the human and environmental dimensions of economic growth; these are typically posed as complements, or adjustments, to such economic measures such as GDP.

REFERENCES

Diamond, J. 2005. *Collapse: How Societies Choose to Fail or Succeed.* New York: Viking Press.
Hunt, T., and C. Lipo. 2011. *The Statues That Walked: Unraveling the Mystery of Easter Island.* New York: Free Press.

Contributors

EDITORS

Jennifer Chirico, PhD, MPH, is the CEO and founder of Sustainable Pacific Consulting, a Certified B Corporation that specializes in sustainability and responsible business. She is an adjunct professor at Hawai'i Pacific University in the Global Leadership and Sustainable Development master's program. Previously, she was the executive director of the Sustainable Living Institute of Maui at UHMC and worked as a senior consultant for Booz Allen Hamilton. She holds a PhD in environmental policy, a master's degree in public health, and a bachelor of science degree in management.

Gregory S. Farley, MS, is the director of the Center for Leadership in Environmental Education at Chesapeake College in Wye Mills, Maryland. He recently served as a visiting scholar with the Sustainable Living Institute of Maui. He is also an associate professor of biology at Chesapeake College, where he has held the Stuart M. Bounds Distinguished Teaching Chair and is the cochair of the college's sustainability efforts. A marine zoologist and evolutionary biologist, he holds a master of science degree in biological science from Florida State University and a bachelor of science degree in biological science from Duke University.

CONTRIBUTORS

John Bendon, MS, LEED AP, BPI, RESNET HERS EMS, is a leader in Hawai'i in the field of green building and energy efficiency. He is a LEED accredited professional with a specialty in LEED Homes, a Certified Energy Rater under the Residential Energy Services Network, and a Building Performance Institute building analyst. He holds a master's degree in real estate and construction management and a bachelor of arts degree in environmental studies. He managed Maui's first commercial LEED New Construction project, inspected and tested all of Maui's certified LEED Homes projects, and has worked on many LEED projects in Hawai'i. He also teaches classes on green building and LEED across Hawai'i, on the mainland, and internationally.

Linda Cox, PhD, is a community economic development specialist with the College of Tropical Agriculture and Human Resources at the University of Hawai'i at Manoa. An economist by training, she has published widely on topics including agriculture and tourism; the marketing of local agricultural products; management of business, human, and natural resources; and other topics that examine the complex interrelationships among cultural, economic, and ecological sustainability.

John Cusick, PhD, is a geographer on the faculty of the Environmental Center at the University of Hawai'i at Manoa. His research interests include environmental history and education, political ecology, and global tourism. He helped create the Hawai'i Ecotourism Association's Certified Ecotour Operators program and was part of the panel that awarded the first certifications under that program.

Scott Fisher, PhD, is the director of conservation for the Hawaiian Islands Land Trust (HILT), a position he has held since January 2011. From 2003 to 2010 he served as a project manager at HILT's 277-acre Waihe'e Coastal Dunes and Wetlands Refuge, where he led all of the ecological restoration work conducted by HILT. Since 2005, he has also served on the Maui/Lana'i Island Burial Council, and he is a member of the boards of Hawai'i Wetlands Joint Venture, Public Access Trails Hawai'i, and the Friends of Midway Atoll.

Reza Ghorbani, PhD, is assistant professor of mechanical engineering and director of the Renewable Energy Design Laboratory at the University of Hawai'i at Manoa. A former Research Fellow of the Natural Sciences and Engineering Research Council of Canada, he is interested in renewable energy device design and applications, actuators and applied control, dynamics, and robotics, in which he holds five international awards.

Matthew Goyke, AIA, NCARB, LEED AP, is the cofounder of Green Sand Architecture + Sustainability. With a degree in architecture from Arizona State University, Matthew has been working in Hawai'i for over two decades years. Strongly committed to the practice of architecture in the context of people's needs and the environment, Matthew is a LEED accredited professional, NCARB certified, and a registered architect in Hawai'i, California, New York, and Oregon. He has been involved in many LEED projects, two National Green Building Standard projects, and two Living Building Challenge projects.

George Kent, PhD, is a forty-year member of the Department of Political Science at the University of Hawai'i at Manoa, from which he retired in 2010. He now serves on the faculty of the Centre for Peace and Conflict Studies at the University of Sydney, Australia, and on the faculty of the Social Transformation Concentration at Saybrook University in San Francisco. His long and distinguished career in food security and human rights advocacy has included service to the United Nations Food and Agriculture Organization, the East-West Center's Environment and Policy Institute, the governments of the Marshall

Islands, the Philippines, and Thailand, and people on four continents and throughout the Pacific Islands.

Penny Levin, MA, is a conservation planner in the field of ecological restoration and the founder and project coordinator of *E kūpaku ka 'āina*—the Hawai'i Land Restoration Institute, a small nonprofit based in Maui. She is also a taro farmer and a longtime student of indigenous agricultural and land restoration practices in Hawai'i, Southeast Asia, and other environs. She received a graduate degree from the University of Hawai'i at Mānoa and was a Fellow with the Participatory Development Group at the East-West Center. She has been a member of the Taro Security and Purity Task Force since 2008, serving to protect and perpetuate traditional Hawaiian taro cultivars in collaboration with botanical gardens, agriculture stations, taro growers, and cultural practitioners throughout Hawai'i.

Steve Parabicoli is the water-recycling program coordinator for the County of Maui. He has been at the forefront of water-reuse projects in Maui County for almost thirty years. He has led the implementation of multiple water projects that have helped Maui County move toward greater water-resource conservation.

Kirk Surry, MS, is the garden coordinator for Kihei Elementary School. He is also a writer, permaculturist, and advocate for healthy nutrition in schools.

Ramsay Remigius Mahealani Taum is a leader in sustainable practices based on Native Hawaiian and indigenous cultures. He is recognized locally, nationally, and internationally for transformational stewardship principles and practices. He is sought after as a keynote speaker, lecturer, trainer, and facilitator. He is especially effective working with Hawai'i's travel, leisure, and retail industry, where he integrates Native Hawaiian cultural values and principles into contemporary business. He received the 2008 East-West Center Leadership Certificate Program's Transformational Leadership in Sustainability award, was recognized in Hawai'i Home + Remodeling, and was named "Who's Keeping Hawai'i Green" 2008 Individual Educator Honoree in *Honolulu Magazine* and *Hawaii Business Magazine*. Mentored and trained by respected *kūpuna* (elders), he is a practitioner and instructor of several Native Hawaiian practices, including *ho'oponopono* (stress release and mediation), *lomi haha* (body alignment), and *Kaihewalu Lua* (Hawaiian combat/battle art). He is a graduate of the Kamehameha Schools, attended the US Air Force Academy at Colorado Springs, and earned a bachelor of science degree in public administration from the University of Southern California.

Shanah Trevenna, MS, was named by *Hawaii Business Magazine* as one of five that will shape Hawai'i for the next fifty years. She is a leading sustainability educator, author, and consultant. Her consulting firm, Smart Sustainability Consulting, is a highly recognized model business for the new economy. She currently teaches in the Honors Program at the University of Hawai'i at Manoa, is the president of the Sustainability Association of Hawaii, and speaks nationally on her book, *Surfing Tsunamis of Change: A Handbook for Change Agents,* which received College Educator International's Sustainability Champion Award for its innovative framework for implementing change.

Luis Vega, PhD, is the manager of the US Department of Energy's National Marine Renewable Energy Center, which is housed at the University of Hawai'i at Manoa. He is also on the faculty of the Hawai'i Natural Energy Institute, also at University of Hawai'i–Manoa. A specialist in wave energy and ocean thermal energy conversion, he has extensive experience in the theoretical design and at-sea testing of ocean energy devices, and he has worked to bring reliable, renewable wind energy to Fiji.

Lauren C. Roth Venu, MS, is the founder and president of Roth Ecological Design. She began her work in applied ecological design under the aegis of John Todd, a renowned pioneer and inventor of ecologically engineered technologies. She and Todd brought the Living Machine technology to Hawai'i under a federal biotechnology program that served as the state's first bioremediation initiative. She founded Roth Ecological Design International, LLC to build the Asia–Pacific Region business. She holds a bachelor of arts degree in environmental science with a specialty in water and biology from the University of Colorado at Boulder and a master of science degree in oceanography from the University of Hawai'i at Manoa.

Susan Wyche, PhD, is currently the special projects coordinator at University of Hawai'i Maui College. At Washington State University, where she directed the Department of English Writing Program, she taught environmental literature courses and proposed the first service-learning-based environmental education program. Later, she studied landscape design at the University of California–Berkeley extension program and for ten years owned a landscape design firm in California specializing in sustainable landscape design. As a board member for the Homeless Garden Project in Santa Cruz, she initiated an edible nursery project. Now living on Maui, she has designed gardens for Maui Nui Botanical Gardens and Kihei Elementary School, and she serves on the advisory board for the school garden organization Grow Some Good.

Index

brackish groundwater, 149
breadfruit, 36, 66. See also *'ulu*
BREEAM, 177. *See also* Building Research
   Establishment's Environmental
   Assessment Method
Brundtland Commission, 60–61
building commissioning, 185–189;
   agents, 185
Building Research Establishment's
   Environmental Assessment Method,
   177. *See also* BREEAM
Bureau of Ocean Energy Management,
   Regulation, and Enforcement, 162

"canoe plants," 47, 65–66
Captain George Vancouver, 49
Captain James Cook, 14, 17, 48, 50
carbon footprint, 231
carrying capacity, 92, 201, 232
CASBEE, 178. *See also* Comprehensive
   Assessment System for Building
   Environmental Efficiency
cattle, 50, 119
Center for Ecoliteracy, 253
Center for Smart Building and
   Community Design, 229
chlorine disinfection, 144, 146
closed-loop system, 252–253; cycle, 231
coagulation, 146, 148, 153
Collective School Garden Network, 253
Comprehensive Assessment System for
   Building Environmental Efficiency,
   178. *See also* CASBEE
Comprehensive Water Reuse Incentive
   Program, 139
Continental worldview, viii, 25. *See also*
   Western worldview; Eurocentric
   worldview; European worldview;
   Western mindset; Western worldview
Cook, Captain James, 14, 17, 48, 50
Coral Reef Alliance, 211–212.
crayfish, 54, 59
crop diversity, 101–102
cultural tourism, 204

DBEDT, 42, 201, 207, 208. *See also*
   Hawai'i Department of Business,
   Economic Development, and Tourism

delamping, 225–228, 232
disaster, 28, 34–37
dryland taro fields, 88, 117; burning of,
   105. See also *kuaiwi*
dumpster dive, 223, 231

Earned Income Tax Credit, 41
Easter Island, 258
ecological design, 125–127, 130
ecological footprint, 231
ecotourism, 199–214
edible schoolyard, 235, 247
effluent: disposal, 146; filtration, 148
endemic species, 49
energy audit, 232
energy efficiency strategies, 184, 185
EnergyStar, 180, 184, 185
Engineers without Borders, 229
Environmental Impact Statement,
   162
Etheridge, Chef Brian, 243
Eurocentric worldview, viii
European worldview, 15
Evolution Sage, 231
Ewa Beach,134

Family and Individual Self-Sufficiency
   Program, 41
Fans, 183–184, 192; bathroom, 194
Farmer's Livestock Cooperative, 134
farms: size of, 119; economic productivity
   of, 119
Federal Energy Regulatory Commission,
   161–162. *See also* FERC
FERC, 161–162. *See also* Federal Energy
   Regulatory Commission
fertilizer: chemical, 55, 105, 117; cost of,
   70; importing, 104
Fiske, Chef Dan, 243
food security and insecurity, 28
Food Security Council, 42
Food Security Task Force, 42
food self-sufficiency, 28, 50, 60, 63, 79, 80,
   92, 100, 112, 132
food: importing and exporting, 105
foot-candle, 225
Forest Stewardship Council, 180, 193.
   See also FSC

wetland taro cultivation, 47, 52, 54, 57, 58; protection of, 69. See also *lo'i kalo*.

WIC, 40. *See also* Special Supplemental Nutrition Program for Women, Infants, and Children

windows, placement, size, and quantity, 182

World Tourism Organization, 199, 206, 208

World Vegetable Center, 253

World Wildlife Fund, 60

Writing Across the Curriculum, 230. *See also* WAC

WWRD, 143–157. *See also* Wastewater Reclamation Division

WWRF, 143–157. *See also* Wastewater Reclamation Facilities

zero waste, 231